DECISION TOOLS FOR PEST MANAGEMENT

DECISION TOOLS FOR PEST MANAGEMENT

Edited by

G.A. Norton

*Cooperative Research Centre for
Tropical Pest Management,
University of Queensland,
Australia*

and

J.D. Mumford

*Department of Biology,
Imperial College at Silwood Park,
Ascot, UK*

CAB INTERNATIONAL

CAB INTERNATIONAL
Wallingford
Oxon OX10 8DE
UK

Tel: Wallingford (0491) 832111
Telex: 847964 (COMAGG G)
Telecom Gold/Dialcom: 84: CAU001
Fax: (0491) 833508

A catalogue entry for this book is available from the British
Library

ISBN 0 85198 783 4 ✓

Typeset by Alden Multimedia, Northampton
Printed and bound in the UK at the University Press, Cambridge

Contents

Contributors

N.D. Barlow;   *Agresearch, New Zealand Pastoral Agricultural Research Institute Ltd, Ellesmere Junction Road, PO Box 60, Lincoln, New Zealand.*

R.K. Day;   *International Institute of Biological Control, PO Box 76250, Nairobi, Kenya.*

C. Garforth;   *Agricultural Extension and Rural Development Department, No. 3 Earley Gate, University of Reading, Whiteknights Road, Reading RG6 2AL, UK.*

C.J. Hamilton;   *Library Services Centre, CAB International, Silwood Park, Buckhurst Road, Ascot SL5 7TA, UK.*

J. Holt;   *Natural Resources Institute, Central Avenue, Chatham Maritime, Chatham ME4 4TB, UK.*

D.L. Johnson;   *Research Station, Agriculture Canada, Lethbridge, Alberta T1J 4B1, Canada.*

T.H. Jones;   *Department of Biology, Imperial College at Silwood Park, Ascot SL5 7PY, UK.*

J.D. Knight;   *Department of Biology, Imperial College at Silwood Park, Ascot SL5 7PY, UK.*

J.D. Mumford;   *Department of Biology, Imperial College at Silwood Park, Ascot SL5 7PY, UK.*

G.A. Norton;   *Cooperative Research Centre for Tropical Pest Management, University of Queensland, Brisbane, QLD 4072, Australia*

J.K. Waage;   *International Institute of Biological Control, Silwood Park, Buckhurst Road, Ascot SL5 7TA, UK.*

Preface

Integrated Pest Management (IPM) is a scientific paradigm (Perkins, 1982) that is now of global significance. Its basic concern is with designing and implementing pest management practices that meet the goals of farmers, consumers and governments in reducing pest losses while, at the same time, safeguarding against the longer term risks of environmental pollution, hazards to human health, and reduced agricultural sustainability.

While the philosophy and ideas of IPM are now widely accepted in the political and scientific arena, the practical implementation of IPM has proved far more difficult to achieve. A number of reasons can be suggested, including:

- Insufficient extension resources to service the needs of farmers who wish to employ IPM (Brader, 1979).
- The need for more emphasis on farmer training to get the IPM message across (Kenmore *et al.*, 1987).
- Commercial pressures on farmers to use pesticides, and the idea that pesticide companies disrupt IPM research and implementation activities (van den Bosch, 1978).

While these factors can be important in specific cases, we believe that a more important reason for a lack of IPM adoption is that the IPM 'product' is often not appropriate to farmers' needs (Norton, 1976; Goodell, 1984). Farmers may perceive IPM as too complex, too expensive, too risky, and just not appropriate to their farming system.

Thus, a different approach is required, which focuses much more clearly on the problem. The basic philosophy that underlies the whole of this book is that we need to spend far more effort defining the problem, and only then search for appropriate solutions. In the current research environment, beset with severe competition for funds, the promotion of novel pest management solutions, such as genetically engineered crops, insect pathogens, and other biotechnological advances, predominates. While it would be foolish to suggest that research into these areas should not continue, it also needs to be recognized that the chances of such options being successfully implemented in the near future are unknown. A much safer bet, particularly in developing countries, is to concentrate effort on employing to better effect the pest management techniques currently at our disposal.

This problem-based approach, with its close resemblance to certain business analysis methods, takes a broad perspective of pest problems, viewing them as only one component in agricultural production. It also requires an inter-disciplinary approach, that encompasses the political, social, and economic dimensions of pest problems, as well as their agronomic, ecological and technological features. Our basic premise is that the adoption of this approach will significantly increase the chance of successful implementation of improved pest management. This will be the case whether the prime concern is with the design of R & D programmes for novel techniques and practices, the development of pest management recommendations, changes to relevant policies, or with improved training.

The decision tools described in this book offer a way of putting this idea into practice, by helping to define problems in a rigorous, explicit, and structured way; by providing a means of systematically searching for acceptable improvements in pest management; and by delivering them to users. Consequently, the prime audience includes research scientists, practitioners, students, and indeed all those who have an interest in the improvement of pest management decision making.

Further, since pest management is one particular example of resource management, having an impact on productivity, sustainability, and environmental and human health hazards, *Decision Tools for Pest Management* offers principles and techniques that are equally applicable to other aspects of resource management. Indeed, we would suggest that the book should be of interest to anyone concerned with the problems of matching applied biology to practical resource management problems, but particularly those involved in the design of strategic research priorities or in the delivery of research results.

Before describing the structure of the book in more detail, we first need to make clear its overall purpose. *Decision Tools* does not attempt to provide a review of the world literature on pest models. Rather, it attempts to provide a working manual for those who are in any way concerned with improving pest management. Consequently, the origins of the book lie in a number of collaborative projects in which the authors have been personally involved, as well as in a number of different training courses, where the authors have attempted to distill their experience and relay the key messages to others. The merit of using these examples is that we can discuss not only the technical aspects of particular tools but can also demonstrate how they have been of practical value when applied to a wide range of real pest situations. In this we believe the book is unique.

One consequence of this approach is that some decision tool techniques probably do not receive the attention they should, since they have not featured large in our experience; this may possibly reflect on our view of their practical utility. Nevertheless, we have attempted to cover the major tools available and, where we have not had first-hand experience ourselves, we refer to the work of other authors. On the other hand, a feature that is included in most of the chapters, is a strong emphasis on economic and decision-making ideas and tools and, for this reason alone, we would maintain that the breadth offered by this book is much greater than other books on the subject.

Another feature of the book, that reflects its origins, is that the philosophy and the approach, as well as the techniques themselves, have been developed over a number of years, in association with many colleagues. It is appropriate at this point, therefore, to thank the following, for the part they have played in shaping the ideas of *Decision Tools for Pest Management*:

- Colleagues with whom we have worked at Imperial College, particularly Michael Way, Richard Southwood, Gordon Conway, Roger Day, Johnson Holt, Jonathan Knight, Hefin Jones, Gareth Edwards-Jones, Martin Birley, Jeff Waage, Tim Denne, Michael Cammell, Nigel Barlow, Hugh Comins, Jim Parlour, David Wareing.
- Other colleagues, with whom we have had many valuable discussions over the years, including Peter Kenmore, Peter Ooi, Lim Guan Soon, Frank Hall, Ian Noble, Terry Stewart, Nick Carter.
- Our many collaborators on specific projects, in particular, Brian Walker, Bob Sutherst, Robin Wilkin, Cheng Jiaan, K.L. Heong, Asna B. Othman, Ho Nai Kin, David E. Evans, John Furlong, Mark Tatchell, Roger Pech, Peter Fenemore, Trevor Lewis, Andrew King, Stuart Smith, John Perfect, Anthea Cook, Julia Compton, Gunter Maywald, Huib van Hamburg, Peter Rawlings, Richard Harrington, Rosa Paiva.
- Many students, including: MSc students in Pest Management at Silwood Park, particularly Trevor Lawson, Bill Griffiths, Colin Taylor, Morag Webb and Moses Kairo; PhD students, particularly Jusoh Mamat, Andy Lane, Tony Smith, Mike Loevinsohn, Antonio Mexia, Lori Peacock, David Gilbert, John Stonehouse, Vania Pivello, Conrad Warwick, Martin Waller and Mark Cox; participants on the two short courses, 'Decision Tools for Pest Management', as well as Peter Combey, who played an important role in organizing the courses, and who introduced the term 'Decision Tools'.
- Participants in research workshops run in the UK and overseas, especially the series of national workshops for the IPM Research Network coordinated by the International Rice Research Institute in the Philippines, under the direction of K.L. Heong.

The 17 chapters in the book can be divided into four sections:

Introduction – Chapter 1 introduces the philosophy, concepts and techniques that lie behind the whole decision tools approach to pest management

Problem definition tools – Chapters 2 to 5 present a range of decision tools that we believe are essential to employ in any pest management project. The techniques covered include primary or conceptual tools, that are aimed at clearly defining the pest management problem, and how to run workshops in which these tools are most effectively used. This section ends with a chapter on survey techniques, describing ways in which information can be obtained on the important issues of how pest managers are dealing with their pest problems, and the factors affecting this decision-making process.

Computer tools – Chapters 6 to 15 present the reader with a range of computer tools that can be used for a variety of purposes. Following an introductory chapter, separate chapters deal with analytical, simulation, rule-based and spreadsheet models. This is followed by chapters on expert systems, pest management games, database systems, geographical information systems (GIS) and pest information retrieval systems.

Implementation tools – The final two chapters are concerned with the use of decision tools in the implementation of pest management; first, the use of decision tools in biological control and then a chapter on the tools generally available for extension of pest management information.

Geoff Norton,   *John Mumford,*
Cooperative Research Centre for Tropical   *Department of Biology*
*   Pest Management*   *Imperial College at Silwood Park,*
University of Queensland,   *Ascot, UK*
Brisbane,
Australia

REFERENCES

Brader, L. (1979) Integrated pest control in the developing world. *Annual Review of Entomology* 24, 225–254.

Goodell, G. (1984) Challenges to international pest management research and extension in the Third World: do we really want IPM to work? *Bulletin of the Entomological Society of America* 30, 18–26.

Kenmore, P.E., Litsinger, J.A., Banday, J.P., Santiago, A.C. and Salac, M.M. (1987) Philippine rice farmers and insecticides: thirty years of growing dependency and new options for change. In: Tait, J. and Napompeth, B. (eds) *Management of Pests and Pesticides: Farmers' Perceptions and Practices.* Westview Studies in Insect Biology, Boulder, Colorado, pp. 98–108.

Norton, G.A. (1976) Analysis of decision making in crop protection. *Agroecosystems* 3, 27–44.

Perkins, J.H. (1982) *Insects, Experts, and the Insecticide Crisis.* Plenum Press, New York, 304 pp.

van den Bosch, R. (1978) *The Pesticide Conspiracy.* Doubleday, Garden City, New York, 226 pp.

1 Philosophy, Concepts and Techniques

G.A. NORTON

Cooperative Research Centre for Tropical Pest Management, University of Queensland, Brisbane, QLD 4072, Australia

INTRODUCTION

This chapter presents a personal view of the philosophy, concepts and techniques associated with the use of decision tools in pest management. It draws very heavily on previous publications (Norton, 1976, 1982, 1987a, 1991; Norton and Mumford, 1983; Norton and Pech, 1988). The chapter consists of five sections:

- General approach and philosophy
- Key concepts
- Techniques
- Case study – rice pests
- Summary

GENERAL APPROACH AND PHILOSOPHY

Research and development in pest management does not always lead to practical improvements. Where this is the case, the problem generally falls within two categories:

1. R & D is aimed at the wrong questions or at developing inappropriate practices. In other words, there is a fault with R & D **design**.
2. Despite R & D being well targeted, the results are not getting through to be implemented by pest managers and their advisors; there is a problem of **delivery**.

What is required, at least in the short and medium term, is for more emphasis to be placed on matching applied science research to the real problem. This idea is by no means new. A number of similar approaches have been proposed for a more problem-oriented approach to other cases of natural resource management (Table 1.1).

However, having the right idea is one thing, putting it into practice is more difficult. Our aim in this book is to show how a range of decision tools can help to implement this

Table 1.1. Other problem-oriented approaches.

- Business management (Cyert and March, 1963)
- Natural hazards research (Kates, 1970)
- Appropriate technology (Schumacher, 1973)
- Village studies (Norman, 1974)
- Agroecosystems analysis (Walker *et al.*, 1978; Conway, 1985)
- Farming systems research (Byerlee *et al.*, 1980)
- Farmer first (Chambers *et al.*, 1989)

approach and to help avoid the two problems mentioned above – poor research design and poor delivery of research results to practical problems. It should be emphasized at this point that our concern here is with applied research strategies. Basic, or curiosity-driven research, although it may well result in novel pest management options in the future, involves different decision criteria. However, if basic research does lead to technological possibilities, then the problem-based approach again becomes relevant.

Part of the problem is that pest problems are complex (Fig. 1.1). They are shaped as much by political, social and economic forces, as by climate, agronomy, technology and

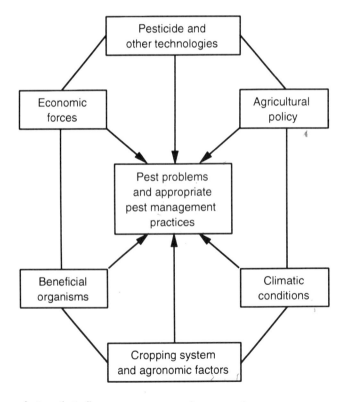

Fig. 1.1. The many factors that affect pest status, pest damage and pest management.

various ecological processes. To fully appreciate and diagnose pest problems and make appropriate recommendations, all the factors shown in Fig. 1.1 need to be taken into account. It should be noted here that we use the term 'pest' throughout this book to include all noxious organisms (insects, diseases, weeds, rodents, birds, etc.) that cause damage to man, livestock, crops and property.

Another problem with implementing improvements in pest management is associated with institutional gaps. Research and extension scientists often operate independently, because of institutional barriers, despite the fact that both have a mutually important role to play in diagnosing problems and designing recommendations.

Finally, and by no means least important, is the need to understand pest management problems from the farmers' point of view; they are, after all, the people most involved in making pest management decisions. We need to appreciate their perceptions of pest problems, how they tackle them, and what on-farm constraints there might be on future improvements.

Decision tools can help to resolve all of these problems by providing an explicit and rigorous means of analysing the decision problems faced by the various players involved in pest management, and facilitating interactions between them (Fig. 1.2).

Before looking in more detail at how this can be achieved, let us first consider a specific example to illustrate the nature of the problem.

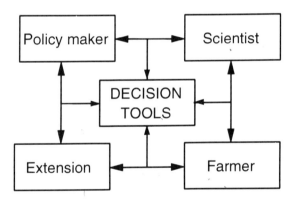

Fig. 1.2. The ways in which decision tools facilitate interactions between participants in the decision-making process.

Case Study – Leaf Cutting Ant in Trinidad

Some years ago I was involved in a project to control the leaf cutting ant, an important pest in South America, and a major problem in citrus, cocoa and other crops in Trinidad (Lewis and Norton, 1973). The traditional method of controlling the pest was to search for ant nests, to mark the nests, and then to treat with pesticide, as a liquid or dust formulation.

In the 1970s, a British government scheme was established to improve ant control by formulating a locally made bait, consisting of citrus pulp and aldrin insecticide, and developing a technique for aerial application. Aerial baiting was field tested on an island off

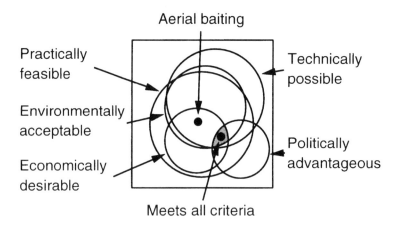

Fig. 1.3. Factors affecting the feasibility, acceptability and desirability of control measures.

Trinidad and found to be at least as effective as the traditional method. Thus, aerial baiting could be classed as **technically possible** (Fig. 1.3).

Since aircraft used for sugar-cane froghopper control in Trinidad would be available for aerial baiting of ants, the technique was **practically feasible**. Aerial baiting was also **economically desirable** as it involves no searching costs and gives economies of scale. Although there appeared to be some human and environmental risk associated with aerial baiting, compared with the traditional practice this risk would be much less. On these grounds, aerial baiting appeared **environmentally acceptable**.

One might conclude from this that aerial baiting was both **feasible** and **desirable**. In practice, however, it was not adopted. Although several reasons can be suggested for this, a major reason for non-adoption appears to be the fact that aerial baiting was not seen to be **politically advantageous**.

The general principle to be learned from this experience is that if a new control practice is to be adopted, it **must** meet all of the criteria shown in Fig. 1.3. If the main aim is to improve pest management, there are two important, practical implications:

R and D targeting – The target area of feasibility, acceptability and desirability (shaded area of Fig. 1.3) has to be identified first; then research and development can be directed towards this target. Lessons can be learned from industry, especially Japanese industry, where customers' needs and rival competition is assessed first, to identify the market niche. R & D programmes are then designed to exploit this niche (see, for example, Ohmae, 1983).

Reducing constraints – The implementation of better pest management can be enhanced if the shaded area of feasibility is increased by reducing key constraints. For instance, this might involve government loan schemes, training programmes, and improved information dissemination.

It is in this context that decision tools can make an important contribution. There are two major roles they can play (Fig. 1.4):

1. Problem specification – decision tools help to rigorously define pest management problems and to identify key questions in the context of farmers' problems. This helps to improve R & D targeting.

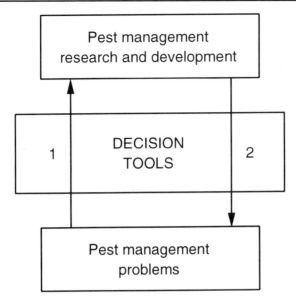

Fig. 1.4. The role of decision tools in pest management.

2. Analysis and delivery – decision tools can be used to assemble research findings, to analyse and interpret them, to determine recommendations, and to provide a means of training and disseminating information.

A later section in this chapter gives examples of the way in which decision tools have been used in these two ways for rice pest management. However, before looking at a practical illustration of the application of decision tool techniques, let us first turn our attention to key concepts that lie behind these techniques, and which are crucial to the whole philosophy and approach being advocated.

KEY CONCEPTS

There are five key concepts which underlie the overall Decision Tools approach – the decision model, information gaps, development pathways, locking-in, and key components and processes. Let us look at each of these in turn, starting with the decision model, the most important concept.

Decision Model

The decisions made by farmers regarding pest management are determined by the four factors shown in Fig. 1.5 (Norton, 1976; Mumford and Norton, 1984):

1. The **problem**, in terms of the pest, the level of attack, and the damage caused.
2. The **options** available to the farmer, such as cultural practices, resistant crop varieties or breeds of livestock, and pesticides.
3. The **farmer's perceptions** of the problem and of the availability and effectiveness of the options.

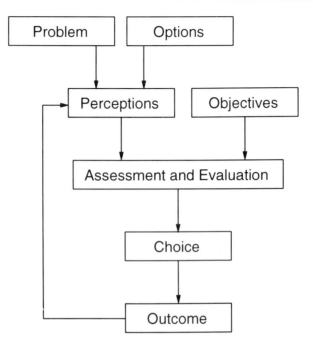

Fig. 1.5. The decision model.

4. The **farmer's objectives**, including monetary goals and his or her attitude to financial risks, health hazards and community values.

We need to understand how the variables in this decision model (Fig. 1.5) influence the actual pest control decisions made by producers, if our research and extension effort is most likely to succeed. Thus, interviews with farmers are often concerned with determining the actual control practices that they carry out, and why – for instance, are they aware of alternatives, how do they make their decisions, and what is their perception of the potential loss caused by the pest?

This decision model can also provide a valuable framework for looking at the decision problems that face policy makers, extension agents and research scientists. For instance, consider the questions that need to be asked in designing an R & D strategy for plant breeding. In deciding on a particular strategy aimed at producing a variety that incorporates disease resistance, rather than one of a number of other possible features, the plant breeder needs to consider a number of factors (Fig. 1.6). This includes an assessment of the likely adoption of the variety if it is successfully produced, requiring consideration of the factors that affect why farmers choose the particular varieties and disease control practices they employ at present. Thus, a key question that arises is – 'What features will the new variety need to have in order to be a feasible and desirable option that farmers will adopt?' This provides a specified target at which the breeder needs to aim.

Thus, in attempting to identify this target, analysis of the decisions made by farmers may reveal that they are constrained from implementing particular control actions by a lack of labour or finance, or by their perception that certain options will increase risk.

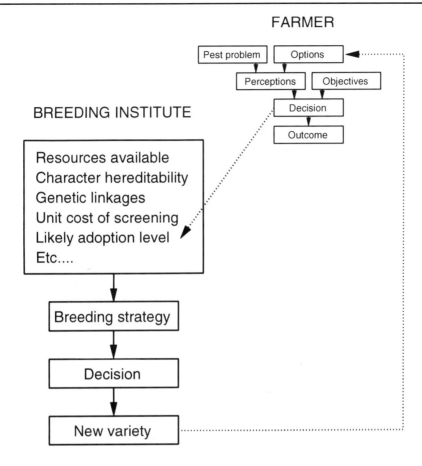

Fig. 1.6. Factors that need to be considered in the design of a crop breeding strategy

A second illustration of the importance of understanding the decision makers' situation in designing R & D strategies is provided by the case of supervised insect pest control in apples.

Case study – Apple pest management in the United Kingdom

The idea behind the strategy of supervised control of insect pests is that growers monitor insect population levels and only spray when the pest population is sufficiently high to justify the cost of treatment. While this seems an eminently reasonable strategy, in practice, the four major constraints shown in Fig. 1.7 become apparent when the growers' situation is studied more closely (Fenemore and Norton, 1985).

It seems likely that it is for these reasons that there has been limited implementation of the recommended supervised control scheme for insect pests: most growers calendar spray. If this type of analysis had been carried out at an early stage, perhaps more appropriate improvements in pest management might have been sought.

Another reason why 'improved' pest management may not be implemented is a lack of knowledge or information. This brings us to the second concept.

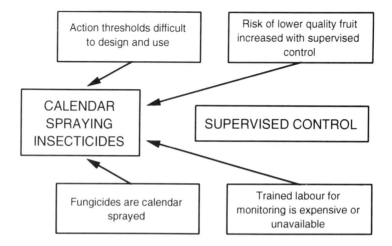

Fig. 1.7. Some factors which lock apple growers into calendar spraying.

Information Gaps

If farmers are to manage pests in the best way, there is a certain set of knowledge and information they need to be aware of, including conceptual and technical knowledge, as well as the 'know-how' to carry out certain practices. There are a number of reasons why farmers may not be aware of this information: for instance, because they are growing a new crop.

Case study – Oilseed rape in the United Kingdom

With the entry of the United Kingdom into the European Community in the early 1970s, the high intervention price for oilseed rape encouraged many cereal growers to grow rape as a break crop, resulting in a dramatic increase in crop area. As pest problems were encountered, this increase in area was followed by an increase in pesticide use (Fig. 1.8). Clearly, this raises two questions – what information should these new growers of oilseed rape be aware of in order to make sensible decisions on the use of pesticides? and, to what extent are they aware of this information?

The answers to the first question are given in Table 1.2. In attempting to answer the

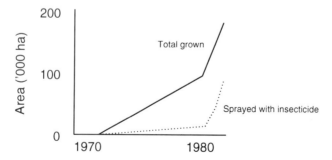

Fig. 1.8. Changes in oilseed rape after United Kingdom entry into the European Community.

Table 1.2. Key information on oilseed rape pest management of which growers should be aware.

- Names of the four most important insect pests:
  Cabbage stem flea beetle
  Pollen beetle
  Seed weevil
  Pod midge

- Recognition of the two insects most prevalent at flowering (pollen beetle and seed weevil)

- Appreciation that pod midge is mainly a headland pest

- Appreciation that the exit hole bored in the pod by the emerging seed weevil larva can be used as an entry point by the pod midge

- Knowledge of the recommended economic threshold for:
  Cabbage stem flea beetle
  Pollen beetle
  Seed weevil

- Appreciation of the ability of the crop to compensate for pollen beetle damage early in flowering

- Appreciation of the possible yield losses that could be caused by the four main insect pests

- Knowledge of the possible harm to bees by spraying during flowering

second question, a survey of growers was carried out to 'test' them on their knowledge of these critical bits of information (Lawson, 1981). It was found that some information, such as the fact that spraying insecticide at flowering can kill bees, was known by all the growers interviewed. On the other hand, pest identification and knowledge of the compensation ability of the crop was only appreciated by some of the growers. Therefore, there was a difference between the information that farmers have and that which they should have, to make good decisions. This is the information gap.

Where an information gap (Norton and Mumford, 1982) exists, we not only need to know how large the gap is but also the reason for this gap. There are several possibilities (Fig. 1.9):

1. Research gap – Some of the required information is just unavailable. Therefore, if the gap is to be closed, appropriate research needs to be carried out.

2. Synthesis/interpretation gap – Research has been carried out on the topic of concern but this information has not been pulled together, analysed and interpreted in the context of the farmer's problem. In this case, appropriate systems and decision analysis tools should be of value.

3. Dissemination gap – Part of the reason for an information gap at the farm level may be that relevant information is just not reaching the farmer. This implies that extension services may need to be improved or that the form in which information is relayed to the farmer needs to be changed.

4. Reception gap – Although information gets through to the farmer, he or she may

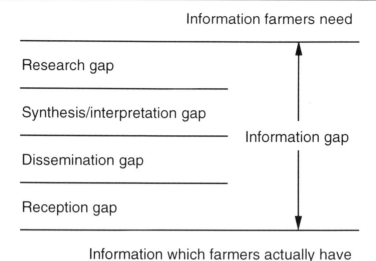

Research gap

Synthesis/interpretation gap
                                                                            Information gap
Dissemination gap

Reception gap

Information which farmers actually have

Fig. 1.9. Information gaps associated with pest management (from Norton and Mumford, 1982).

not be able to utilize it properly because he or she lacks the necessary background knowledge. If this is the case, more effort devoted to farmer training is implicated.

Clearly, if decision making at the farm level is to be improved, effort needs to be devoted to these four activities according to their contribution to the overall information gap. An inevitable problem, however, is that the cause, nature and degree of this information gap is constantly changing, as new pest problems arise and novel means of controlling them are developed. This brings us to the third concept – development pathways.

Development Pathways

Virtually all agricultural systems are changing. Cereal systems in the United Kingdom and rice systems in Asia have become far more intensive in recent years. In parts of Africa, dramatic shifts are taking place from traditional multiple cropping systems to more mechanized, sole cropping systems. Such changes can influence pest management in a number of ways – by increasing the favourability or susceptibility of the crop to pest attack, by changing the effectiveness of control measures, or by altering farmers' objectives.

Agricultural development can be viewed as a dynamic process crossing a landscape. Its direction will depend on the topography of the area which, in turn, will be determined by such factors as political processes, economic forces, and technological change and adoption (Fig. 1.10). The adoption of certain technologies can also lead to the adoption of other, complementary technologies (Crouch, 1981), pushing the producer along a particular development path, which may lead to increasing pest problems and affect the ability of farmers to control them. Thus, changing agricultural practices can lead to a change in the need for information and knowledge on pests and their control at the farm level (the information gap). Since there are inevitable time lags between the initiation of

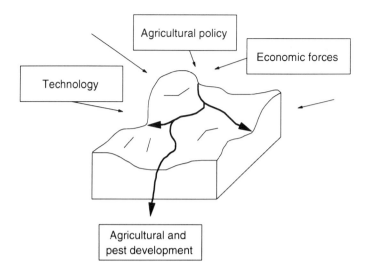

Fig. 1.10. The agricultural development process.

research and the implementation of its results, scientists have to try to predict the direction of these changes to ensure that their research is directed towards producing relevant information and appropriate technology.

To illustrate how agricultural development can have a considerable impact on pest control practices, consider the recent changes that have occurred in sugarbeet production in the United Kingdom.

Case study − Sugarbeet seedling pests in the United Kingdom

Traditionally, sugarbeet was grown by sowing the seed at high density and then hoeing at a later stage to give the optimal density for maximum root yield. As rural labour became scarce and more expensive, sugarbeet growers looked for ways to reduce the labour input. Two technological developments enabled them to do this (Fig. 1.11):

1. Development of the precision drill, which allows sowing to a prescribed density.
2. Breeding of monogerm beet seed.

Thus, in combination, these two developments allowed growers to sow to the optimal stand, eliminating the need for hoeing (Fig. 1.11). However, this development also produced a more susceptible crop. Any thinning by seedling pests, which previously was compensated for by adjacent plants in the densely sown crop, now caused a reduction in yield for every plant lost (Fig. 1.11). The less dense stand of seedlings also made the crop more attractive to aphids, which transmit serious virus diseases. Thus, the next wave of technology to be adopted was granular insecticide, used in a routine manner to protect this more susceptible crop (Fig. 1.12).

There are two practical implications we can draw from the effect that agricultural development can have on pest problems and pest control. First, but by far the most difficult to implement, is the idea that we can attempt to divert agricultural development along a different pathway, that has less, long-term, disruptive effects. The second impli-

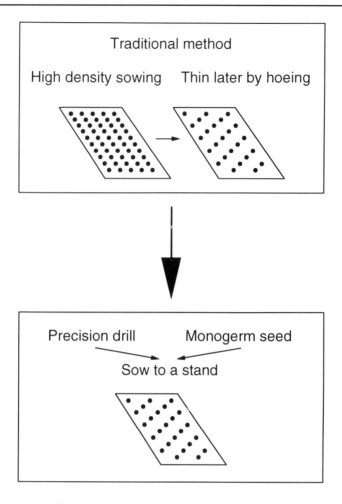

Fig. 1.11. Changes in sugarbeet growing in the United Kingdom.

cation is that we can at least be prepared for likely increases in pest problems, and initiate or re-direct research strategies accordingly. As we will see subsequently, an historical profile is one means of systematically pulling together relevant information on system development and stimulating ideas on likely development scenarios and their implications for future pest management problems.

A major feature associated with many development pathways is the loss of flexibility that can occur, resulting in one particular type of development becoming locked-in. This is the subject of the fourth key concept.

Locking-in

Several of the case studies described above have illustrated the idea of 'locking-in'. In the case of apple IPM, we saw that growers are locked-in to a calendar treatment strategy by a number of factors, preventing them from easily switching to an action threshold strat-

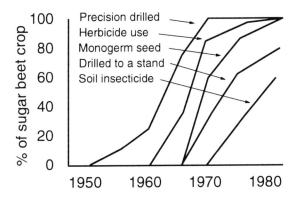

Fig. 1.12. Adoption of new technology and practice of growing sugarbeet in the United Kingdom (after Dunning, 1982)

egy. Similarly, sugarbeet growers have become locked-in on a particular development path, the result of adopting two technologies, that enabled them to reduce labour costs. Once farmers had made the initial investment to adopt the precision drill, for use in a range of crops, they were beginning to get locked-in. With the subsequent development and adoption of monogerm seed, growers became even more locked-in, with the result that they had to resort to insecticides to deal with the pest problems that ensued.

In a similar way, government programmes can become locked-in on particular strategies. Throughout the world there are a number of pest eradication and vaccination programmes that have become locked-in for a variety of reasons:

- Because of large initial investments in capital equipment, such as breeding houses for flies in area-wide, sterile male release programmes.
- For institutional reasons, where the prime objective becomes institutional survival rather than pest control, *per se.*
- Because there are serious political risks in attempting to change the strategy.

Scientists can also get locked-in on particular research strategies, their decisions on R & D programmes being influenced directly or indirectly by dominant science paradigms.

Table 1.3. Paradigms in agricultural entomology in the USA (paraphrased from Perkins, 1982).

Paradigm 1 – Chemical control
  'The major tool for controlling insects is the application of toxic chemicals to them'

Paradigm 2 – Integrated pest management (IPM)
  'Because pest organisms are capable of countering control efforts (e.g. pesticide resistance), we must understand nature's methods of regulating populations and maximise their application' (C.B. Huffaker)

Paradigm 3 – Total Pest Management (TPM)
  'I have great confidence in the ingenuity of young scientists to perfect the technology necessary to put sound principles of insect suppression into practice ... reducing total populations on an ecosystem basis in an organised and coordinated way' (E.F. Knipling)

John Perkins' (1982) study of the history of agricultural entomology in the USA identified three paradigms (Table 1.3) that have had a significant influence in their time.

The practical significance of a particular scientific paradigm being dominant is that it will determine the projects supported by funding agencies. In the USA, the UK and many other countries, the research being carried out is moulded by the IPM paradigm. The research questions asked and the approaches adopted are 'pre-conditioned' by the IPM philosophy. Undoubtedly, this is appropriate for certain situations but not all, as we have seen for the strategy of supervised control of insect pests in apples.

One of the major objectives of *Decision Tools* is to develop a more objective approach to the problem of improving pest management. It attempts to avoid preconceived notions, based on predominant paradigms, of how improvements can best be achieved. The paradigm we would like to instil is more problem based. The philosophy of this approach is to first define the pest problem, then to specify the constraints to improvement, and only then look for appropriate solutions. The fifth key concept is related to this idea, that the key components and processes affecting the problem need to be understood first.

Key Components and Processes

At the beginning of this chapter the importance of trying to understand the full dimensions of pest problems was stressed. However, when tackling specific pest problems, it is impossible to investigate or account for all of the factors that might influence the development of pest populations, the damage they cause, or their control. We must necessarily simplify. The question is – by simplifying, surely we will miss important features of the problem? Here is where this fifth concept is of relevance since a fundamental premise on which the Decision Tools approach is based is that the cause of pest problems, and the opportunities for resolving them, lie in a few key components and processes.

The need to identify key components and processes will be a recurring theme throughout the remainder of the book. The purpose of most of the primary decision tool techniques described in the next chapter is to help in simplifying in the right way, and not to end up with a distorted simplification of the problem. This 'reductionist' idea also lies behind most modelling approaches. A good example is provided by Noble and Slatyer (1980), who determined a limited set of key processes – or 'Vital Attributes' – that determine the way in which particular plant species respond to climatic and management perturbations they experience in a particular environment. They used a rule-based model to simulate ecological developments. For insect and disease pests key factors, or vital attributes, that affect their development and control, include temperature thresholds for development, generation times, mobility of the pest, genetic resistance to pesticides or adaptation to host resistance and reproductive potential.

TECHNIQUES

Two roles of decision tool techniques were described earlier, in relation to Fig. 1.4: problem specification, and analysis and delivery. Let us now look in more detail at the way in which decision tools can be used in practice.

Problem specification — Decision tools can be used to systematically define the pest management problems experienced by a range of decision makers and so determine current constraints, identify research gaps, and assess the feasibility of 'novel' methods of control. In this way, decision tools can highlight those issues on which future policy, research, development and extension strategies should be targeted, to maximize the chance of successful implementation.

Analysis and delivery — Decision tools can also be used to pull together research findings and expertise, to interpret them in the context of farmers' problems, to design pest management strategies, and to provide specific recommendations in keeping with the design specifications. In addition, they can also be used to provide a means of training and disseminating information.

In this context, the decision tools themselves, used to perform these two roles, can be classified into two groups:

Primary decision tools — including techniques such as flow charts, historical and seasonal profiles, interaction matrices, and decision trees. Primary decision tools are used to help identify the key components and relationships associated with pest problems and so contribute to the identification of key questions, opportunities and constraints, that relate to the improvement of pest management. Consequently, these techniques are mainly concerned with problem specification, and the design of strategies for dealing with the problem. This also includes the specification of secondary decision tools.

Secondary decision tools — including database systems, simulation models, expert systems, and other, more detailed, decision tools, which are often, but not always, computer based. Secondary decision tools are mainly used for analysis and delivery, being concerned with the design of control recommendations, providing pest forecasts, or delivering specific information and advice. In addition, secondary decision tools can also provide a means of clarifying the problem, and specifying more clearly what needs to be done.

Therefore, in practice, a clear distinction between primary and secondary decision tools, and the different functions of problem specification, analysis and delivery, cannot be made. For instance, during the development of expert systems, which are usually thought of as delivery tools, important gaps in information are frequently highlighted, giving a precise specification of the research that needs to be done to improve pest management advice. However, it is also true that in developing expert systems, primary decision tools are often used as a means of acquiring the information and knowledge to be incorporated as rules in the expert system.

Thus, the pragmatic definition of decision tools used here distinguishes between those techniques (primary tools) which should be employed for all pest problems, as opposed to those (secondary tools) which are only appropriate for certain pest problems. Indeed, as indicated above, the use of primary tools should determine the need for secondary tools and provide a clear specification for their design.

The following two chapters describe primary decision tools in detail, and later chapters describe a range of secondary decision tools. In the meantime, to give some idea of the way in which a series of decision tools can be employed in practice, we now turn to the specific problem of rice pest management.

Case Study – Rice Pests

Rice pests can be viewed as components in a hierarchy of nested systems, ranging from international and national systems down to the village, farm, paddy and hill. When considering the genetic features of rice pests, for example, and the potential for pesticide-resistant strains to move from one area to another, the large-scale migration of pests on climatic fronts has to be viewed in the context of an international system. However, to appreciate details of predator/prey interactions, that can affect the level of insect pest attack, processes operating at the level of the paddy or the hill need to be investigated.

Thus, when undertaking research into pest problems and designing recommendations for control, **relevant** factors in a number of systems may need to be considered. The stress on **relevant** is important since it is the major purpose of **primary decision tool techniques** to help determine which are the relevant or key factors that need to be considered for any particular problem.

This point can be illustrated by considering rice pest problems in Malaysia (Norton and Heong, 1988). A **primary decision tool** – in this case, a flow chart (Fig. 1.13) – indicates the major factors, at a national, regional and farm level that can influence the overall changes in pest status.

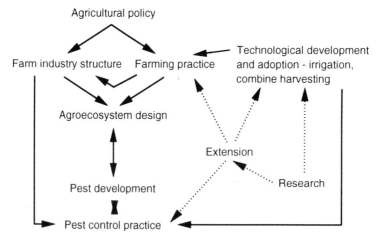

Fig. 1.13. National and lower system level interactions in Malaysia. · · · · · information flow.

For instance, agricultural policy in Malaysia has included providing a guaranteed price for rice, subsidized fertilizers and investment in irrigation schemes. The net effect of this has been to increase the ability and incentive of farmers to improve rice production while, at the same time, reducing the risks of adopting the necessary technology. These factors, along with others, notably the increasing shortage of rural labour, have contributed towards the dramatic changes in rice production practices that have been observed in the major rice growing area of Muda in recent years.

Table 1.4. Expert opinion on the impact of agronomic change on rice pest status, compared with traditional practice (after Norton and Heong, 1988).

	Cultural practices				
	Direct seeding		Synchronous planting	Stubble burning	Double cropping
Rice pests	Broadcast	Drilled			
Leaf folders	+	?	− −	?	+
Stem borers	+ +	+	−	− − −	0
Green leafhoppers/ tungro	+ +	?	− −	− −	+ +
Planthoppers	+ + +	+ +	− −	−	+ + +
Bugs	+ +	?	− −	− −	+ +
Rats	+ +	?	−	−	+ +
Blast	+	?	+ +	− −	+ +
Sheath blight	+	?	− −	− −	+ +
Weeds	+ +	+	?	−	0

+, + +, + + +: slight, moderate, considerable increase in pest attack, respectively, thought likely to occur; −, − −, − − − : slight, moderate, considerable reduction in pest attack, respectively, thought likely to occur; 0: no significant effect thought likely; ?: 'don't know'.

Clearly, the changes in the agroecological features of the crop associated with these developments are likely to have serious implications for pests. One way to investigate this is to carry out field trials and farm surveys. However, this will involve considerable research resources and, in the case of the latter, can only be undertaken once the changes are well established. Although field work is essential, a preliminary, and low-cost means of obtaining some indication of the likely changes that might be expected in pest status were obtained by seeking the opinions of experts who had been working on rice pests for more than 5 years, including scientists, extension and agricultural officers (Norton and Heong, 1988). A summary of the results of this exercise is given in Table 1.4.

Another important aspect of this initial analysis of rice pest systems, using **primary decision tools**, is the analysis of decision makers, including those concerned with pest control policy as well as the rice farmer himself. In both cases, decision makers are attempting to respond to pest problems in a way that will best meet their objectives. However, such decisions are often made under highly constrained circumstances, such as limited capital, equipment and 'know-how'. If resources are to be allocated to those activities that have most impact in improving pest management, then an understanding of these decision making processes and their 'failings' is essential.

The following example illustrates how farmer surveys can contribute to improved research and extension. A survey of 915 rice farmers in Malaysia, to investigate sprayer usage (Heong *et al.*, 1992), identified urgent needs for improving knapsack sprayers, training farmers in sprayer techniques, and improving the quality of locally produced knapsack sprayers. This study led to the initiation of a new research programme in pesticide application technology and to the extension service establishing 'sprayer clinics' and centres

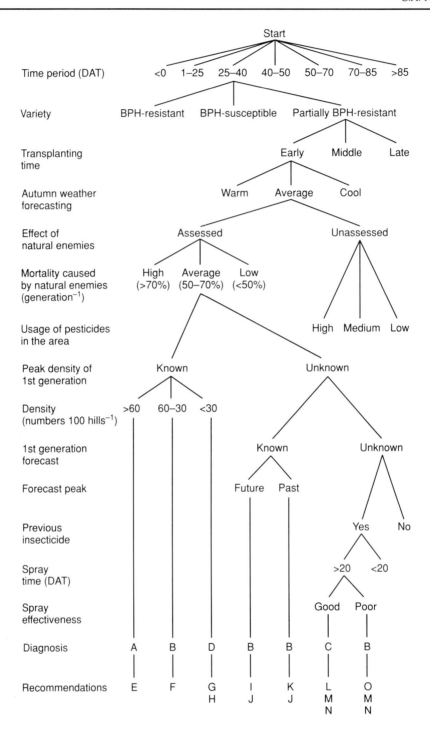

Fig. 1.14. An outline of the expert system for BPH control in Zhejiang Province, China (Holt et al., 1990)

for providing advice, simple repairs, training demonstrations, and discussions on how to improve spraying techniques.

So far, we have concentrated on the analysis of rice pest problems in general. To illustrate how **primary** and **secondary** decision tools can be of value in analysing specific pest problems, we now turn to one particular pest – the brown planthopper – that has been, and still is, of major concern throughout the rice growing areas of Asia.

Brown Planthopper (BPH)

Viewed in terms of the hierarchy of systems described earlier, there are features of this particular pest problem that have to be considered at all levels. Since BPH is a migratory pest, capable of long distance movement on prevailing winds, populations in one country or region can contribute both genetically (in terms of pesticide resistance genes, or genes that confer adaptation to 'resistant' rice varieties) as well as numerically to populations in other countries or regions.

To investigate the problem of BPH within the context of the damage they can cause to particular rice crops, and farmers' options for controlling this pest, the system that is of most relevance is the paddy or rice field. In this case, migration of BPH, which obviously can play an important role in determining the risk of populations developing to damaging levels, is an input from a 'higher' system level.

A number of 'key' questions, relating to the development and control of BPH populations, have been investigated using a **secondary decision tool** – in this case, simulation modelling. However, prior to the construction of this simulation model, a **primary decision tool**, an interaction matrix, was of value in structuring the system, and helping to identify the key components that needed to be included in the model. Further details on this can be found in Chapter 6.

Once a BPH simulation model had been constructed, it could be used for two major purposes – to improve our understanding of BPH population biology, particularly the causes of outbreaks, and to identify good control strategies. A powerful attribute of computer simulation models, which is often the main reason for their development, is the ability to reproduce scenarios and test hypotheses to a much greater extent than is possible experimentally. In a Chinese version of the BPH model, the performance of insecticide control measures has been investigated across a wide range of situations. The main purpose has been to determine 'robust' strategies, which are strategies that give an acceptable performance for a wide range of circumstances.

The lessons learned from these 'simulation' experiments, together with field data and field experience, have been incorporated in another **secondary decision tool**, an expert system. An outline of this system is provided in Fig. 1.14. The expert system is currently being used to establish links between research and extension scientists, providing a mutual learning experience.

SUMMARY

This book is concerned with looking at decision tool techniques in detail, the principles involved and how they can be **usefully** employed for improving decision making in policy, research, and advice in pest management. They can be of help in:

- Precisely defining pest problems
- Assessing the feasibility of potential control methods and practices
- Specifying the need for information at the farmer level
- Identifying key research and advisory questions
- Generating hypotheses about the reasons for pest outbreaks or increases in pest status
- Evaluating the technical and economic performance of control options
- Determining recommendations for specific situations

REFERENCES

Byerlee, D., Collinson, M., Perrin, R., Winkelmann, D. and Biggs, S. (1980) *Planning Technologies Appropriate to Farmers – Concepts and Procedures.* Centro Internacionale de Mejoramiento de Maiz y Trigo, Mexico.

Chambers, R., Pacey, A. and Thrupp, L.A. (1989) *Farmer First.* Intermediate Technology Publications, London.

Conway, G.R. (1985) Agroecosystem analysis. *Agricultural Administration* 20, 31–55.

Crouch, B.R. (1981) Innovation and farm development: a multi-dimensional model. In: Crouch, B.R. and Chamala, S. (eds) *Extension Education and Rural Development Vol. 1.* John Wiley and Sons, pp. 119–134.

Cyert, R.M. and March, J.G. (1963) *A Behavioural Theory of the Firm.* Prentice-Hall, Englewood Cliffs, New Jersey.

Dunning, R.A. (1982) Sugar beet pest, disease and weed control and the problems posed by changes in husbandry. In: Austin, R.G. (ed.) *Decision Making in the Practice of Crop Protection.* British Crop Protection Council, Croydon, pp. 91–98.

Fenemore, P.G. and Norton, G.A. (1985) Problems of implementing improvements in pest control, a case study of apples in the U.K. *Crop Protection* 4, 51–70.

Heong, K.L., Jusoh, M.M., Ho, N.K. and Anas, A.N. (1992) Sprayer usage among rice farmers in the Muda area, Malaysia. In: Lum, K.Y. (ed.) *Pesticides in Tropical Agriculture.* Malaysian Plant Protection Society, Kuala Lumpur. pp. 327–330.

Holt, J., Cheng, J.A. and Norton, G.A. (1990) A systems analysis approach to brown planthopper control on rice in Zhejiang Province, China. III. An expert system for making recommendations. *Journal of Applied Ecology* 27, 113-122.

Kates, R.W. (1970) *Natural Hazards in Human Ecological Perspective: Hypotheses and Models.* Working Paper no. 14, Natural Hazard Research, University of Toronto.

Lawson, T.J. (1981) Information flow in pest management, with reference to winter oilseed rape. Unpublished MSc thesis, University of London.

Lewis, T. and Norton, G.A. (1973) Aerial baiting to control leaf-cutting ants (Formicidae, Attini) in Trinidad. III. Economic implications. *Bulletin of Entomological Research* 63, 289–303.

Mumford, J.D. and Norton, G.A. (1984) Economics of decision making in pest management. *Annual Review of Entomology* 29, 157–174.

Noble, I.R. and Slatyer, R.O. (1980) The use of vital attributes to predict successional changes in plant communities subject to recurrent disturbances. *Vegetatio* 43, 5–21.

Norman, D.W. (1974) Rationalising mixed cropping under indigenous conditions: the example of Northern Nigeria. *Journal of Development Studies* 11, 3–21.

Norton, G.A. (1976) Analysis of decision making in crop protection. *Agroecosystems* 3, 27–44.

Norton, G.A. (1982) A decision analysis approach to integrated pest control. *Crop Protection* 1, 147–64.

Norton, G.A. (1987a) *Pest Management and World Agriculture: Policy, Research and Extension.* Papers in Science, Technology and Public Policy, No. 13, Imperial College, UK.

Norton, G.A. (1987b) Developments in expert systems for pest management at Imperial College, U.K. *Review of Marketing and Agricultural Economics* 55, 167-70.

Norton, G.A. (1987c) A strategic research and development approach to improving animal health — especially tick and tick borne diseases. Paper presented at a workshop in Porto Allegre, Brazil, on *Ticks, Tick-borne Diseases and Insect Pests of Cattle in the Southern Cone Countries of South America: II Workshop — Progress Since 1983.*

Norton, G.A. (1991) Formulating models for practical purposes. *Aspects of Applied Biology* 26, 69–80.

Norton, G.A. and Heong, K.L. (1988) An approach to improving pest management: rice in Malaysia. *Crop Protection* 7, 84–90.

Norton, G.A. and Mumford, J.D. (1982) Information gaps in pest management. *Proceedings of the International Conference on Plant Protection in the Tropics*, Kuala Lumpur, Malaysia, 1–4 March, 1982, pp. 589–597.

Norton, G.A. and Mumford, J.D. (1983) Decision making in pest control. *Advances in Applied Biology* 8, 87–119.

Norton, G.A. and Pech, R.P. (eds) (1988) *Vertebrate Pest Management in Australia: a Decision Analysis/Systems Analysis Approach.* CSIRO — Division of Wildlife and Ecology, Canberra.

Norton, G.A. and Walker, B.H. (1985) A decision analysis approach to savanna management. *Journal of Environmental Management* 21, 15–31.

Ohmae, K. (1983) *The Mind of the Strategist: Business Planning for Competitive Advantage.* Penguin Books. 283 pp.

Perkins, J.H. (1982) *Insects, Experts, and the Insecticide Crisis.* Plenum Press, New York.

Schumacher, E.F. (1973) *Small is Beautiful: a Study of Economics as if People Mattered.* Blond and Briggs Ltd, London.

Walker, B.H., Norton, G.A., Conway, G.R., Comins, H.N. and Birley, M. (1978) A procedure for multidisciplinary eco-system research based on the South African Savanna Eco-system Project. *Journal of Applied Ecology* 15, 481–502.

2

Descriptive Techniques

G.A. NORTON[1] AND J.D. MUMFORD[2]

[1]Cooperative Research Centre for Tropical Pest Management, University of Queensland, Brisbane, QLD 4072, Australia: [2]Department of Biology, Imperial College at Silwood Park, Ascot SL5 7PY, UK

INTRODUCTION

In the previous chapter, we discussed the philosophy and concepts of decision tools for pest management. A major conclusion from this was the need to understand the nature of pest problems before attempting to improve them. In this chapter we introduce some of the primary or conceptual techniques that can be used to undertake such analyses.

The major purpose in employing primary decision tool techniques is to produce a **systematic, directed** and **focused** definition of the problem, and particularly to identify the key components and processes affecting pest status, pest damage and pest control. As discussed in Chapter 1, this provides a basis for targeting research, development and advisory effort on the key questions associated with feasible and acceptable pest management options.

Since pest problems, like other natural hazards, arise through the interaction between natural and human use systems, the definition of pest problems must necessarily involve components of these two systems. The first major task of problem definition in pest management, therefore, is system description. The scope of this description and the level of detail that is required will be determined by the problem and the particular questions being asked. For instance, how has the pest problem arisen? How is it likely to change in the future? What are the opportunities and constraints to improved pest management in the future?

The second task of problem definition, and one which raises a number of further questions, is the investigation of the specific decision problems associated with the pest problem. Decision analysis provides a framework and techniques of value in seeking answers to such questions as – What are the policy options for improving pest management? What critical research needs to be carried out before improvements in pest management can be achieved? If farmers are to be trained, what concepts, techniques and information do they need?

Although the two tasks of pest problem definition (system description and decision analysis) are very closely connected, for the purpose of discussing relevant techniques

Table 2.1. Outline of analyses conducted in the Coronel Pacheco workshop, Brazil, December, 1988 (Norton and Evans, 1989).

Level	Techniques	Purpose
Regional level	Historical profile Discussion with dairy scientists	To identify regional trends and indicate likely changes in the tick problem
Farm level	Five case studies Discussion with advisers and farmers	To identify the main tick problems faced by farmers and advisers
	Life cycle for brainstorming	To identify tick control options
	Interaction matrix	To identify biological links and interactions
	Information and knowledge needs of farmers and advisers	To provide a target for training courses
	Advisory system model	To provide the basis for an expert system

they are dealt with separately. In this chapter we focus on primary decision tools concerned with system description techniques. In Chapter 3 we concentrate on decision analysis techniques. However, before looking at these techniques in more detail, two important points need to be made.

First, in both chapters it should be borne in mind that the techniques described are not exhaustive but are simply illustrative of a range of available tools; the analysis of a specific pest management problem is unlikely to require the use of all the techniques. Once the objectives of a particular analysis are set, the members of the analytical team choose or discard techniques to meet those objectives as they see fit. In choosing examples to illustrate primary decision tool techniques we have either deliberately chosen simple examples, to illustrate the points quickly, or chosen well-developed examples, to indicate their application to real problems.

The second point is that primary or conceptual decision tool techniques are usually most effective when employed in well-focused, multi-disciplinary workshop groups. To illustrate how they have been used in practice, Table 2.1 shows the techniques employed, and the purpose that each technique was intended to serve, during a workshop in Brazil on tick and tick-borne diseases.

SYSTEM DESCRIPTION TECHNIQUES

As noted in Chapter 1, pest management problems are complex. The development of pest populations, the damage they cause and the way they are controlled, are affected by a range of influences, including biological, ecological, technical, economic and political components and processes. These components and processes are part of a hierarchy of systems, as shown in Fig. 2.1. Thus, a particular pest problem can be affected by policy decisions, taken at a national system level, concerning

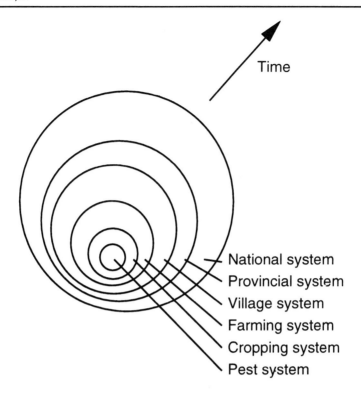

Fig. 2.1. Hierarchy of systems associated with pest management problems.

crop price or pesticide regulation, by agroecological factors, at the farming system level, or by physiological or predation/parasitism processes at the cropping system level. Note that interactions between these components and processes not only occur at a particular system level but can also occur between levels. A policy change to increase crop price, for instance, can result in farmers increasing the inputs of fertilizer used on the crop, making it more susceptible to pest attack and, therefore, causing an increased use of pesticide.

Another important dimension of pest problems, implied in Fig. 2.1, is the fact that they are changing over time, as changes occur in various key factors and processes at different levels in the hierarchy of systems. These changes might be the result of political, economic, or technological developments, as discussed in the previous chapter, or more associated with ecological or biological developments, such as the development of pesticide resistance in a pest population.

Since most of us have been trained as specialists in a particular discipline, our ability to cope with this complexity is limited. We have been trained to analyse specific components in detail and do not have the tools to deal with complex, multi-disciplinary problems (Waddington, 1977). It is precisely to meet this need that the following descriptive tools are being presented. The major purpose is to provide a rigorous and systematic means of determining the key components and processes which influence pest problems, how these factors are changing over time and how they influence the way in which pest management can be improved. Three important classes of technique are considered in this chapter – flow charts, time profiles and matrices.

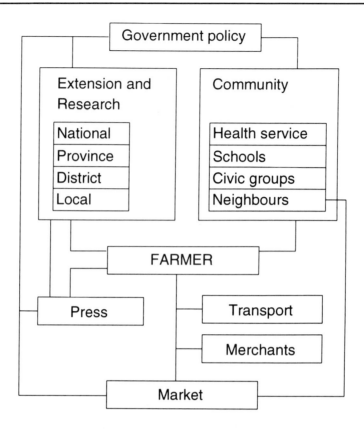

Fig. 2.2. A flow chart illustrating human participants in a farming system.

Flow Charts

Flow charts are among the simplest descriptive tools. They can be used to think about and describe the boundaries of the system, and the relationships between components of a pest problem.

Figure 2.2 sets out the relationships between human participants which could influence the decisions made by a farmer concerning his farming system. The lines indicate important, direct linkages. Subsystems can be shown by boundary lines within the overall system, such as the boundary enclosing the public extension hierarchy. The type of linkage between the different players in this system could also be identified; for instance, an economic link with merchants, an information link with extension agents and a regulation link with civic groups.

In Fig. 2.3, a qualitative element is introduced into a flow chart, describing part of a fruit fly system. Arrows show the direction of each relationship, what is affecting and what is affected. The '+' and '−' signs indicate whether the affected component is increased or decreased by the relationship. This form of flow chart is useful in illustrating indirect relationships. For example, in Fig. 2.3, the indirect relationship between removing soil litter and flies could well be negative since there are two '+'s' and one '−' along the path of direct relationships.

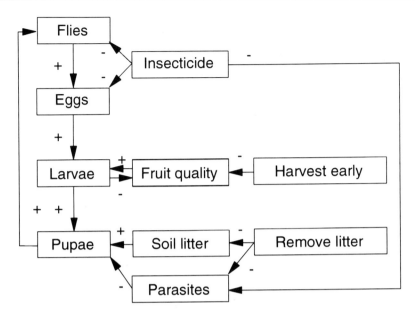

Fig. 2.3. A flow chart with qualitative relationships between some of the components in a fruit fly system.

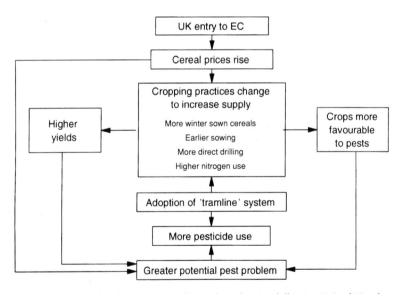

Fig. 2.4. Flow chart showing the development of cereal production following United Kingdom entry into the European Community (after Norton, 1986).

Flow charts can also be used to describe agricultural development, such as the changes in cereal growing in the United Kingdom following entry into the European Community in 1973 (Fig 2.4).

Simple graphic presentations like this, as well as providing an interactive tool for workshop discussion, can help to determine the structure of the problem, and to provide a basis for thinking about the key components and processes, within various systems, that

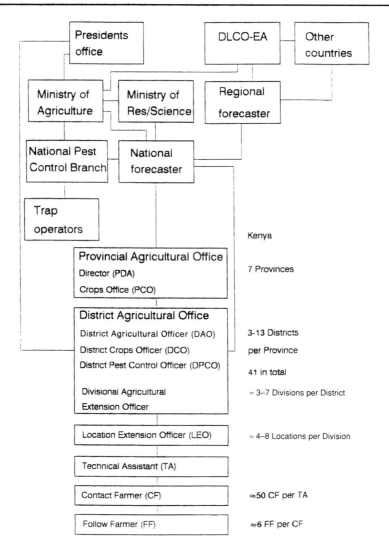

Fig. 2.5. Major players and organizations involved in decision making for East African armyworm control in Kenya (after Day, unpublished). (DLCO-EA = Desert Locust Control Organization – East Africa.)

can influence the pest problem. This can then be used in various ways: for instance, to provide a focus for dividing the components into workable units for detailed research and development.

Organization chart

Another type of flow chart is an organization chart, that indicates the links between various players within and between organizations. The specific example given in Fig. 2.5 shows how this technique was used as part of the initial specification process for the development of a database management system for the East African armyworm (described in Chapter 13). It was important to gain an appreciation of the major players or participants in the system, and the flow of information between them. The organization chart shown in Fig. 2.5, in describing the major players and the relationships between them, helped to identify the District Agricultural Officers as the major clients for the output of the database system, and led to interviews with them to determine the information that would be most useful to them.

Time Profiles

As well as determining the relationships between key components and processes, the time dimension of pest problems also needs to be investigated if a meaningful description of the problem is to be obtained. Two types of time profile, described below, are particularly important – the historical and the seasonal profile.

Historical profiles

To understand the broad dimensions of pest problems, and so help identify constraints and opportunities for resolving them, it is necessary to see current problems in terms of an historical perspective. This general idea was discussed in Chapter 1, when we considered the concept of development pathways. A technique to help provide this perspective is the historical profile.

The first step in constructing a historical profile is to identify the major factors that directly or indirectly influence the development of the pest problem. However, note that additional factors can always be added as the analysis proceeds and reveals further key factors. The next step is to determine the relevant time period over which changes in these factors are to be considered; 20 years or more is often desirable. However, this decision will clearly depend on the particular situation. A series of graphs can then be drawn, using expert opinion available in the workshop group or, where a more detailed analysis is being made, from various statistical sources. The important point to emphasize here is that the level of detail required in drawing the historical profile is simply that which is sufficient to indicate important trends and relationships. Further detail, apart from being time consuming to collect, may add little to a good, rough assessment.

Two examples illustrate the value of an historical profile constructed during an interdisciplinary workshop session. The first example is taken from a workshop, carried out in Brazil in 1986, to make a rapid appraisal of tick and other pest problems on cattle in Rio Grande do Sul. To structure the discussion, the axes shown in Fig. 2.6 were drawn. The participants were then asked to indicate the trends for each of the items, providing quantitative bench-marks wherever possible. A simplified version of the results of this half-day session is given in the figure.

This example illustrates well the various roles that such a technique can play:

- It provides a structured means of bringing together information on a range of aspects that may all have had some influence on the development of the problem.
- Used as a workshop tool, the historical profile sets an agenda, helps to focus discussion, provides a simplified method of communication between participants from different disciplines, and stimulates interaction.
- In the process of constructing the historical profile, hypotheses or key questions are

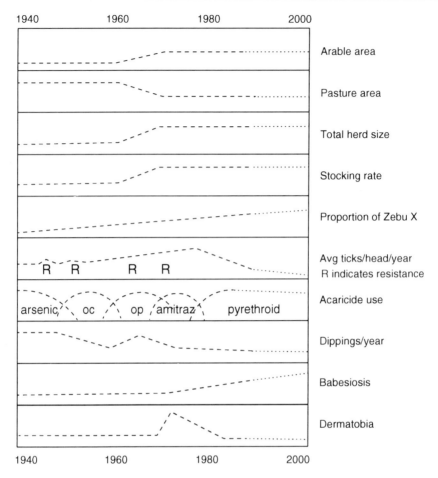

Fig. 2.6. Historical profile of tick and other pest problems on cattle in Rio Grande do Sul, Brazil (after Norton and Evans, 1989).

raised. For instance, in the particular example illustrated in Fig. 2.6, one can hypothesize that the increased incidence of babesiosis (a disease transmitted by ticks) results from the reduction in tick numbers which, in reducing transmission levels, has caused a loss of immunity in young cattle.

- Finally, and perhaps most usefully, the historical profile provides a rigorous basis for thinking about possible future developments (or scenarios) that can affect the status of tick problems and the ability of cattle producers to deal with them.

The second example, shown in Fig. 2.7, is an historical profile for cauliflowers that was constructed during a workshop on vegetable pests in the United Kingdom, held at the Institute of Horticultural Research, Wellesbourne, in 1990. Over the last 40 years many changes have taken place in cauliflower production, harvesting, and marketing. Of most significance for pest management, particularly the management of cabbage root fly, is the change in production methods, from bare root transplanting to the use of peat modules.

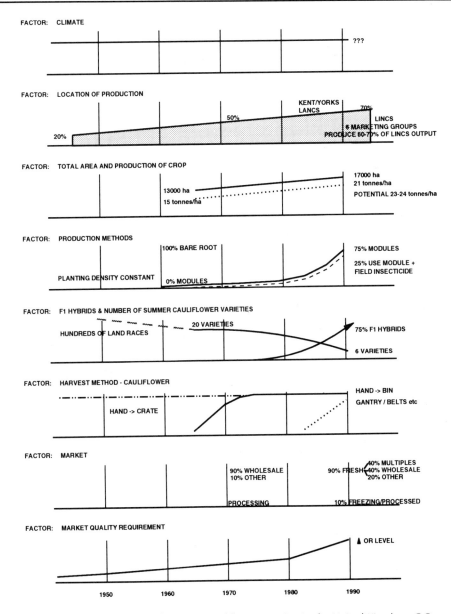

Fig. 2.7. Historical profile of major changes in cauliflower growing in the United Kingdom. OC = organochloride; OP = organophosphate; CRF = cabbage root fly; OSR = oil seed rape.

When growers used the bare root method, they prophylactically treated against cabbage root fly, applying insecticide granules in the soil. Growers still treat prophylactically but, where they use modules, they drench the module in insecticide prior to planting, resulting in a 90% reduction in active ingredient per hectare and a similar cost reduction.

The third, and final, example of an historical profile concerns the large irrigated cotton scheme in the Gezira in Sudan. A major pest that has developed in recent

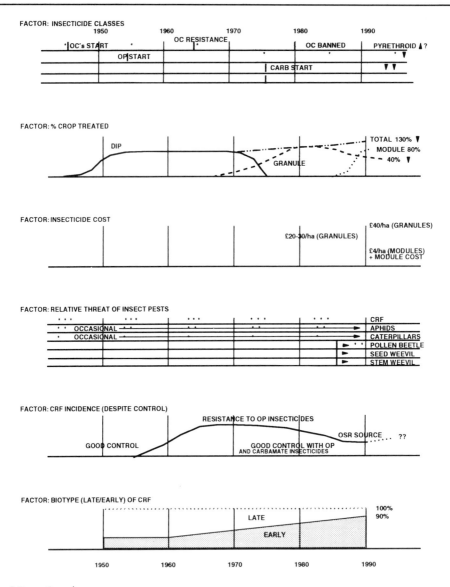

Fig. 2.7. continued.

years is the whitefly (Bemisia). While short term solutions to this problem have been sought, using different insecticides and methods of application, in the longer term, it is important to consider the underlying reason(s) for the increased status of this pest, which could provide a basis for a more sustainable pest management strategy.

The historical profile shown in Fig. 2.8 traces the major developments in the Gezira from 1925 (Griffiths, 1984). Changes that occurred in the area and rotation of crops in the mid-1950s probably made an important contribution to the increase in heliothis and the subsequent increase in the use of insecticide. This is likely to have been one of the major factors contributing to the increase in whitefly. Changes in institutional and financial arrangements have also had an important influence on

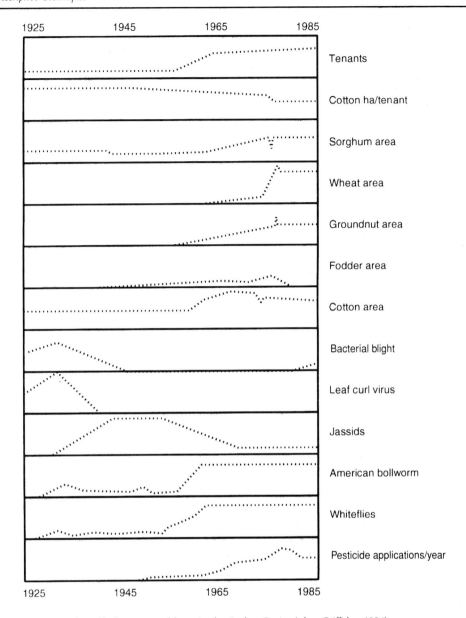

1925   1945   1965   1985

Tenants

Cotton ha/tenant

Sorghum area

Wheat area

Groundnut area

Fodder area

Cotton area

Bacterial blight

Leaf curl virus

Jassids

American bollworm

Whiteflies

Pesticide applications/year

1925   1945   1965   1985

Fig. 2.8. Historical profile for pest problems in the Sudan Gezira (after Griffiths, 1984).

cotton production and protection, by affecting the tenant farmers' incentive to grow good cotton. All these factors need to be considered in developing future strategies for integrated pest management.

Seasonal profile

The seasonal profile is a descriptive technique for considering shorter term changes at the farm or field level, on an annual or crop season basis. As with the historical profile, the first

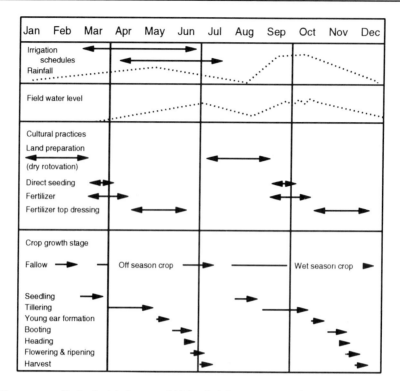

Fig. 2.9. Rice crop profile in the Muda area of Malaysia (after Anon., 1990).

step is to identify the major components involved. The trends in each component can then be graphed over the season.

The examples shown in Figs 2.9 and 2.10 are for rice in Malaysia. To investigate the pest problems faced by farmers in the Muda area in more detail, and to provide a basis for identifying the relationships between the cropping system and rice pests, various seasonal profiles were constructed (Anon., 1990). Initially, a 'standard' cropping profile for the year was constructed, indicating the major cropping practices throughout the season – including land preparation, seeding, fertilizer application, water level and crop growth stages (Fig. 2.9).

Seasonal profiles for individual pests and beneficial agents were then constructed, indicating the development of the pest over the cropping season, and the decision points at various times in the season when control actions could be taken. These 'pest profiles' are shown in Fig. 2.10 for brown planthopper (BPH), tungro, *Echinochloa* and rats. Decision points identified for the management of *Echinochloa* are also shown.

The combination of pest and farming system profiles, including cropping patterns and labour availability, helps scientists from different disciplines to see how the particular pest they are working on fits into the overall system.

Decision profile

Decision profiles are a particular form of seasonal profile. Figure 2.11 shows a decision profile for one particular pest, the wheat bulb fly in the United Kingdom. In addition

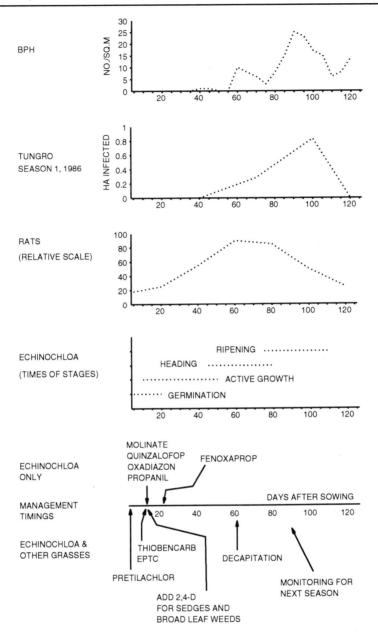

Fig. 2.10. Rice pest profiles in the Muda area of Malaysia (after Anon., 1990).

to a profile of the pest life cycle, Fig. 2.11 shows the points in the season when control decisions can be made and the time at which information is potentially available to the farmer in making these decisions. This particular seasonal profile was used as a means of specifying the structure of an expert system for assessing the risk of wheat bulb fly attack and providing recommendations for control (Jones *et al.*, 1990).

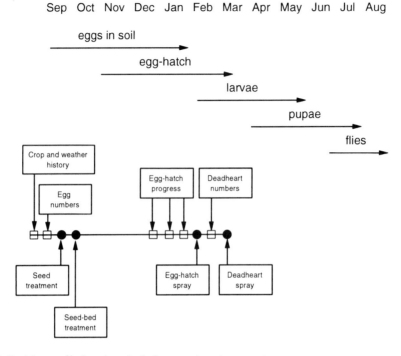

Fig. 2.11. Decision profile for wheat bulb fly control in the United Kingdom (after Jones *et al.*, 1990). Squares represent information inputs and solid circles are decision points.

Matrices

This third type of descriptive technique provides an alternative to flow diagrams as a means of setting out the relationships between key components that affect a particular pest problem. The advantage of the matrix technique is that it provides a more structured approach and forces those completing the matrix to consider each potential interaction. It is usually best to use a flow chart first, to help identify the key components and/or processes to include in the matrix. Let us now look at a number of applications of the matrix approach to different aspects of pest problems.

Conflict matrix

Decision making for many so-called pest problems is often made more difficult by the fact that the organism concerned is not seen by all the players as a pest (Mumford and Norton, 1987). Indeed, it may be seen as a valuable resource by some. A well-known example in Australia is the weed 'Patterson's curse' (*Echium plantagineum*) of dryland cereals. To bee-keepers and graziers this same plant is known as 'Salvation Jane', because of its ability to provide sustenance for bees and cattle when other sources are not available. Clearly, biological control of 'Patterson's curse' could result in costs to bee-keepers and graziers.

Another Australian example illustrates the use of a conflict matrix to clarify the

	Farmers	Amateur hunters	Commercial harvesters	State Agric Depts	State Conservation Depts	Dept Primary Industry	Environmentalists	Animal welfare groups
Eradication	●			?	●		●	●
More control								
Status quo						?		
Enhancement		●	●					

Fig. 2.12. Likely attitudes of different interest groups to four policy objectives concerning feral pigs in Australia. (Enhancement includes increased habitat, distribution and density.)

political tensions that are likely to exist where different groups place a positive and negative value on the same organism. Feral (wild) pigs in Australia are a pest problem to certain crop growers, present potential disease problems to pig farmers and to cattle ranchers (because feral pigs could transmit Foot and Mouth Disease if it is introduced into Australia) and are seen as undesirable by conservationists. On the other hand, amateur hunters and commercial harvesters see feral pigs as a valuable and exploitable resource. These conflicts can be shown by constructing a matrix that consists of the various interested parties on one axis and the different policy objectives that might be sought on the other (Fig. 2.12). Two immediate conclusions are evident. First, none of the groups are thought to be in favour of maintaining the status quo; most are likely to favour outcomes that are very different. The second point is that there is very clear polarization between two groups; those who would favour eradication and those who would like to see feral pig numbers increase. In undertaking an analytical approach to determining the likely outcomes of different strategies and assessing their costs and benefits, the political dimension, portrayed in Fig. 2.12, clearly needs to be included in any final recommendations.

LaGra (1990) presents another form of conflict matrix which he calls participant analysis. A matrix is drawn up in which the positive and negative impacts of options are listed for target groups, support groups, opposition groups or other affected groups. Once all the positive and negative impacts have been listed explicitly, they form a check list for developing strategies to minimize conflicts.

Damage matrix

Where there is a complex of different pests attacking a crop or a cropping system it may be useful to put pests into categories so that they can be considered as groups for

	Beans					Maize				
	Store	Pods	Leaves	Stem	Roots	Store	Ears	Leaves	Stem	Roots
Scarab grubs										X
Noctuid stalk borers							X	X	X	
Aphids			.					X		
Weevils						X	X			
Primary storage pests	X					X	X			
Secondary storage pests						X				
Root lesion nematode										X
Bean fly			X	X						
Aphids		X	X	X						
Bollworm		X					X			
Weevils			X	X						
Maruca		X								
Root knot nematode					X					

Fig. 2.13. A damage matrix for some pests of maize and beans.

certain purposes, such as design of control strategies or development of research programmes.

A damage matrix classifies pest problems according to the type of damage they cause. As shown in Fig. 2.13, pests attacking maize and beans are listed on the horizontal axis, while the parts of the plants they attack are listed on the vertical axis.

A damage matrix such as Fig. 2.13 allows you to see at a glance which pests attack maize ears, which extend over stems, leaves and pods on beans, and which parts of the two crops suffer the greatest range of pest species. It also shows that only one insect pest (bollworm) attacks both crops.

At this stage it is appropriate to remember the purpose of these techniques. System description is used to identify key components in a process in an explicit manner. The techniques we have considered so far have been concerned with obtaining an overview of system components and processes. Let us now look at more detailed descriptive methods, which take us to a more complex level of system description.

Interaction matrices

The use of an interaction matrix as a means of identifying the basic structure of a pest system, or its 'life system', has proved useful in many projects. The first step is to identify key components in the system. The next is to construct a matrix which allows the primary effect of any column component on any row component to be shown.

	Climate 1-5	Mgt 6-11	Crop 12-21	Other organisms					BPH																		
				22-26 Other pests	27-30 Natural enemies	31 Viruses	32 Diseases	33 Weeds	34 Immigration	35 Movement	36 Mating	37 Oviposition	38 Fecundity	39 Egg density	40 Egg survival	41 Egg development	42 Nymph density	43 Nymph survival	44 Nymph development	45 Wing morph	46 Sex ratio	47 Adult density	48 Adult survival	49 Adult development	50 Emigration
Crop 12-21			●															●				●			
Other organisms																										
Other pests 22-26					●												●	●				●			
Natural enemies 27-30																										
Viruses 31																										
Diseases 32																										
Weeds 33																										
BPH																										
Immigration 34																										
Movement 35																										
Mating 36																										
Oviposition 37			●		●																					
Fecundity 38					●																					
Egg density 39					●																					
Egg survival 40			●		●																					
Egg development 41																										
Nymph density 42			●		●																					
Nymph survival 43			●		●																					
Nymph development 44			●																							
Wing morph 45			●																							
Sex ratio 46																										
Adult density 47					●																					
Adult survival 48																										
Adult development 49			●																							
Emigration 50			●																							

Fig. 2.14. Part of an interaction matrix for the rice brown planthopper (after Holt *et al.*, 1987). Outside the BPH–BPH cell only major interactions are shown and limited detail is given of the factors included in the rows and columns of that cell (the number of factors is shown by the row and column numbers). Within the BPH–BPH cell all interactions are shown (● is a major interaction, ○ is a minor interaction).

In the interaction matrix which was constructed for the brown planthopper (*Nilaparvata lugens*), a major pest of rice (Holt *et al.*, 1987), each dot or square in a cell indicates that that particular column component has a primary effect on the row component (Fig. 2.14). Taking any particular column, the open and solid circles in that column indicate the primary effects the component represented by the column has on the system. Similarly, for any row, the open and solid circles in that row indicate the factors that have a primary effect on that component.

Clearly, interactions can occur between these primary effects and they can also lead to secondary and subsequent effects, as one proceeds through the matrix, generally from the top left-hand side to the bottom right-hand side. For instance, pesticides can have a primary effect on canopy predators and subsequently they have a secondary effect on brown planthopper by modifying the effect of canopy predators on the brown planthopper.

In this particular case, the matrix has been used not only to identify relationships but to help determine which of these need to be included in a simulation model aimed at pulling together ecological information on the brown planthopper. The solid circles in the matrix are those effects that are thought to be of such importance that they have to be included in the simulation model. Note that the lack of solid circles associated with management variables does not mean they are unimportant but that for this particular exercise they were taken as given. In subsequent analysis, as described in Chapter 8, the BPH model has been used to assess the efficacy of different insecticide strategies. Another goal of this project has been to investigate the impact of control practices on brown planthopper populations and to develop models for rice-growing areas other than the Philippines. Thus, as the project continued, further squares have been identified in the left side of the interaction matrix.

A second example of an interaction matrix developed in association with a pest modelling exercise is shown in Fig. 2.15 for the Australian cattle tick (Norton *et al.*, 1984). In this case, the ringed dots and the broad arrows indicate the components and processes that have been included explicitly in the tick population simulation model. This matrix has been useful in relating the general results from the model back to particular sites, where a different soil type, for example, can affect the appropriate survival values included in the model (Fig. 2.15).

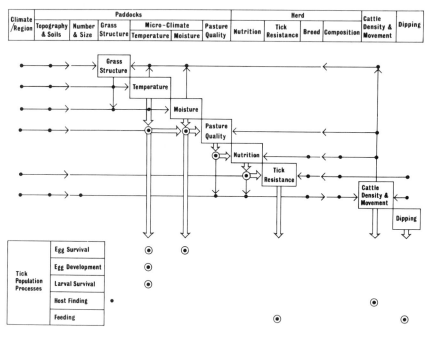

Fig. 2.15. An interaction matrix for the Australian cattle tick (from Norton *et al.*, 1984).

In other projects, interaction matrices have proved useful in planning the design of experimental studies at a laboratory and field scale. In providing a convenient framework for identifying relationships, assessing those likely to be of most importance, and then determining what crucial information is missing, a precise 'research need' is specified. The final example of an interaction matrix, given in Fig. 2.16, is for an important aphid-transmitted virus disease of cereals in Europe, Barley Yellow Dwarf Virus (BYDV). In collaboration with colleagues at the Institute of Arable Crops (Rothamsted) and the Agricultural Development and Advisory Service in Britain this matrix was constructed prior to a two-day workshop. After discussion during the workshop further modifications were made. This matrix, which also has a written commentary for each relationship (an extract is given in Fig. 2.17), together with other techniques, has provided a basis for determining future research and information needs.

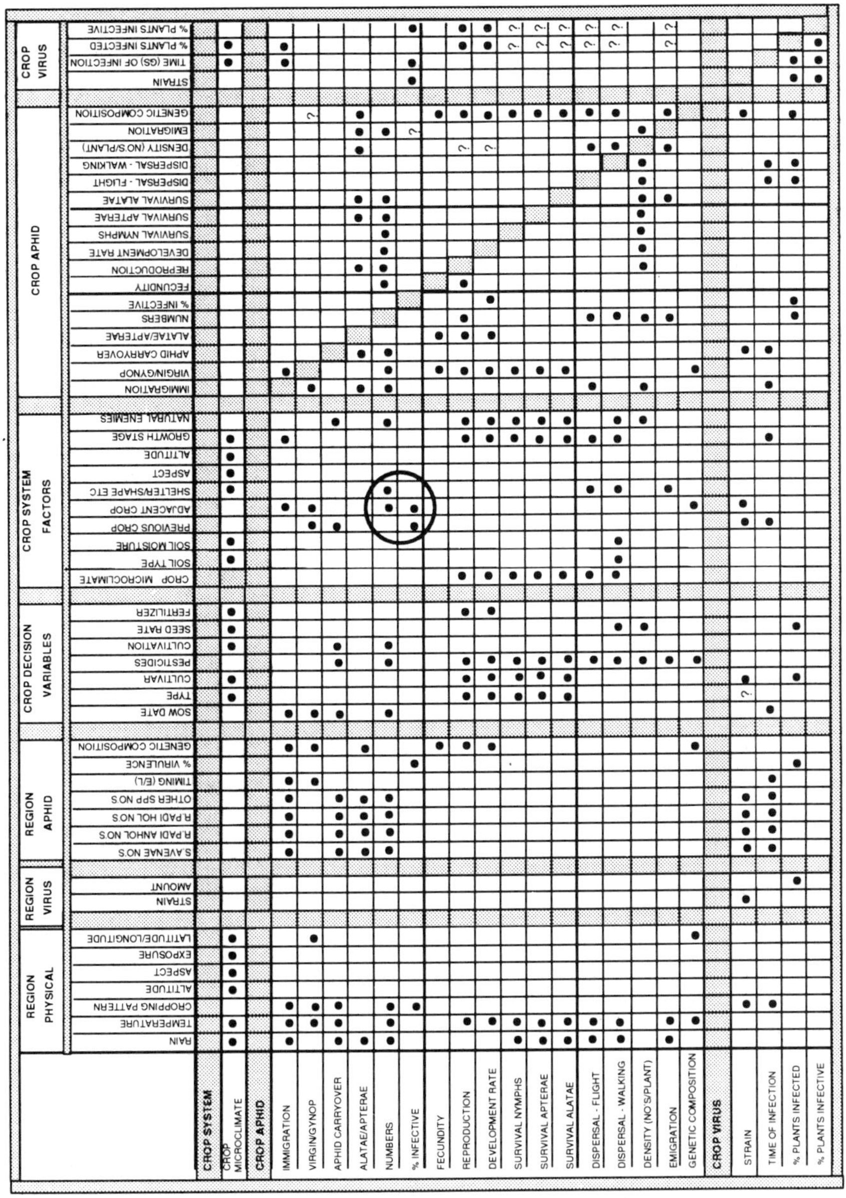

Fig.2.16. Interaction matrix for Barley Yellow Dwarf Virus in the United Kingdom

CROP SYSTEM FACTORS			
		Previous crop	Adjacent crop	Shelter/Shape
C R O P	Numbers		Numbers will be greater since there will be more aphids if the next crop is suitable for cereal aphids	The shape and degree of shelter in a field will determine how many aphids will alight in the field while migrating
A P H I D	% Infective	If the previous crop is a host of BYDV then the number of infective aphids present will be greater	If the neighbouring crop is a reservoir of virus then the number of infective aphids is likely to be higher	

Fig. 2.17. An extract of the commentary on the interaction matrix for BYDV. The four relationships concerned, which are referred to by grid reference, are denoted by a ring in Fig. 2.16.

REFERENCES

Anon. (1990) *Report of a Workshop on Rice Pest Management in the Muda Area of Malaysia*. Department of Agriculture, Kuala Lumpur, Malaysia.

Griffiths, W.T. (1984) A review of the development of cotton pest problems in the Sudan Gezira. Unpublished MSc thesis, University of London.

Holt, J., Cook, A.G., Perfect, T.J. and Norton, G.A. (1987) Simulation analysis of brown planthopper (*Nilaparvata lugens* (Stal.)) population dynamics on rice in the Philippines. *Journal of Applied Ecology* 24, 87–102.

Jones, T.H., Young, J.E.B., Norton, G.A. and Mumford, J.D. (1990) An expert system for the management of *Delia coarctata* (Diptera: Anthomyiidae) in the United Kingdom. *Journal of Economic Entomology* 83, 2065–2072.

LaGra, J. (1990) *A Commodity Systems Assessment Methodology for Problem and Project Identification*. Postharvest Institute for Perishables, College of Agriculture, University of Idaho, Moscow, Idaho, USA.

Mumford, J.D. and Norton, G.A. (1987) Economic aspects of integrated pest management. In: Delucchi, V. (ed.) *Protection integree: quo vadis?*, PARASITIS 86, pp. 397–407.

Norton, G.A. (1976) Analysis of decision making in crop protection. *Agroecosystems* 3, 27–44.

Norton, G.A. (1986) *Pest Management and World Agriculture: Policy, Research and Extension*. Papers in Science, Technology and Public Policy, Number 13. Imperial College, London.

Norton, G.A. (1987) Developments in expert systems for pest management at Imperial College, U.K. *Review of Marketing and Agricultural Economics* 55, 167–170.

Norton, G.A. and Evans, D.E. (1989) *Report on a Series of Workshops on Tick and Tick Borne Disease Control Held in Brazil 1–22 December, 1988*. EMBRAPA, Brasilia, 57 pp.

Norton, G.A. and Heong, K.L. (1988) An approach to improving pest management: rice in Malaysia. *Crop Protection* 7, 84–90.

Norton, G.A., Sutherst, R.W. and Maywald, G.F. (1984) The case of the Australian cattle tick. In: Conway, G.R. (ed.) *Management of Pest and Disease Systems*. Wiley Ltd, pp. 381–394.

Waddington, C.H. (1977) *Tools for Thought*. Jonathan Cape, London.

3 Decision Analysis Techniques

G.A. Norton[1] and J.D. Mumford[2]

[1]Cooperative Research Centre for Tropical Pest Management, University of Queensland, Brisbane, QLD 4072, Australia: [2]Department of Biology, Imperial College at Silwood Park, Ascot SL5 7PY, UK

Introduction

It was suggested in the previous chapter that pest problems are best defined in two ways, using system description techniques and decision analysis techniques. In this chapter we focus on the latter.

When tackling any pest management problem, the importance of analysing relevant decision-making processes cannot be overemphasized. As already indicated in Chapter 1, this does not just mean the ultimate decision maker, such as the farmer, but also includes policy makers and research and extension decision makers. Thus, the decision analysis approach to a pest management problem starts by asking the following questions:

- Who are the decision makers involved?
- How do their decisions affect each other?
- What are the major factors that affect their decisions?

The initial step in this decision analysis approach is a descriptive one, concerned with determining the important 'players' or decision makers in the system and the interactions between them, and identifying the key factors that determine why decision makers currently make the decisions they do. The organization chart and the conflict matrix, described in Chapter 2, give some indication of the techniques that can be used to answer the first two questions.

In describing individual decision makers' problems, the decision model described in Chapter 1 (and repeated in Fig. 3.1) provides the framework for such an analysis. It is important to analyse the problem faced by decision makers in a structured way as early as possible. This can be achieved through exploratory surveys, which, as well as determining what the current practices of decision makers are, will assess:

- Options which are only feasible to decision makers under certain circumstances, and other constraints that might affect the feasibility of options in the future.

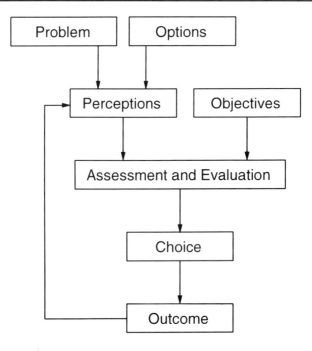

Fig. 3.1. The basic decision model.

- The decision maker's perceptions of the problem in terms of the losses caused by pests, the likelihood of pest outbreaks, the effectiveness of different practices and the decision maker's views on likely future developments.
- The decision maker's economic and non-economic objectives, and his or her attitude to risk.

Details of how one can gather the raw information concerning the factors affecting decision making, through interview surveys, focused group discussion, and other methods, are described in Chapter 5. Here we are concerned with decision tools that can help us to structure the problem and to integrate the information that exists. This not only serves to define the decision maker's problem and identify information and research gaps, but can also lead to a clear specification of what needs to be done to improve pest management, whether it be improved decision support, modifying existing control options to make them more appropriate, or undertaking critical research projects. To provide a framework for the approach, let us first look at the processes involved in pest management decision making and then consider how descriptive tools can help.

Figure 3.2 shows the three major processes involved in making a decision or recommendation on pest management:

- First, a diagnosis of the problem is required, identifying the pest and assessing the level of damage it is likely to cause.
- The second stage is to consider the options for pest management, and to assess their feasibility, cost and effectiveness in reducing pest damage.
- Finally, an assessment of the outcomes associated with each option is undertaken, in terms of the objectives being sought by the decision maker. This will involve some

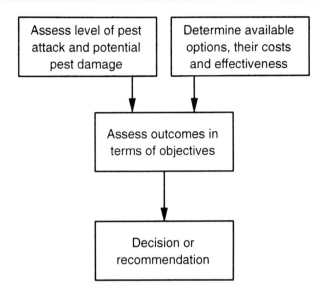

Fig. 3.2. Processes involved in pest management decision making.

form of cost–benefit assessment, as well as other considerations, including safety and environmental concerns. This provides the basis on which the pest management decision or recommendation can be made.

Decision tools can play several roles in contributing to this overall decision-making process. To look in more detail at these roles, let us take each of the three processes described above in turn.

Assessing Potential Pest Damage

The assessment of potential pest damage is important at all decision-making levels. At a political level, the allocation of public funds to research and implementation programmes on specific pests will depend, to some extent, on estimates of losses caused by these pests. This will involve consideration of such factors as the area affected, the levels of loss caused per hectare and the distribution of losses, for example, between large and small farms. Here we focus on the problem faced by the farm decision maker, whose interest is in assessing the potential level of pest loss in order to decide on an appropriate level of pest management response.

Damage Relationships

The damage relationship describes the level of loss associated with different levels of pest attack. Although this relationship is often non-linear, for decision-making purposes it is only necessary to distinguish two forms of damage relationship (Fig. 3.3). A linear damage relationship (Fig. 3.3A) is associated with pests that are vectors of disease, where the crop is unable to compensate, or where the major concern is the presence of

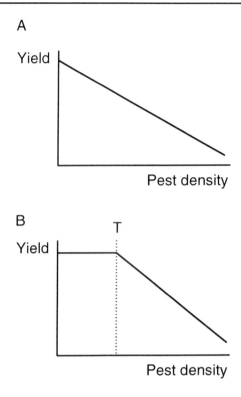

Fig. 3.3. Linear and threshold damage relationships.

a pest causing quarantine or cosmetic 'damage'. In this case, the slope of the damage relationship is the critical factor determining the degree of control required.

The second relationship, threshold damage, has the form shown in Fig. 3.3B. There is
a level of tolerance associated with low levels of pest attack, or compensation to damage
occurs, particularly where this damage is to parts of the plant which are not part of the
marketable yield, such as leaves and roots of fruit crops. In this case, the threshold level
has most effect on the required level of pest management.

For strategic decision making, it is likely to be sufficient to simply determine which
type of damage relationship occurs. For instance, classical or inundative biological control
is unlikely to be successful for a pest causing serious linear damage, since the degree of
control achieved is still likely to result in an unacceptable loss of revenue. On the other
hand, if threshold damage is caused, the damage threshold (T in Fig. 3.3B), below which
the pest does not cause loss in revenue, may be sufficiently high that biological control
agents can keep the population below this level.

To illustrate the value of using these two types of damage relationship for thinking
about the level of information needed, let us turn to a specific, vertebrate pest problem –
rabbits in Australia. In order to decide what investment it might be worth making in
attempting to control rabbits on a property, some information on the damage caused
will clearly be of value.

During a workshop in Australia (Norton and Pech, 1988), factors likely to affect the
type and slope of the damage relationship were discussed. In a season that is drier than
usual, and therefore resulting in less grass, and when the level of stocking is high, any

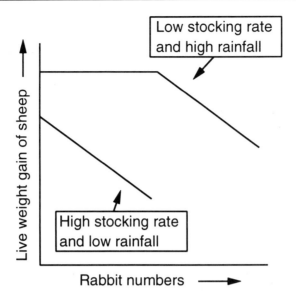

Fig. 3.4. Damage relationships for rabbits in Australia.

vegetation grazed by rabbits is likely to reduce intake by sheep or cattle, resulting in a lower turn-off weight. In other words, there is a linear relationship (Fig. 3.4). The slope of this line can be assessed very crudely in terms of the number of rabbit equivalents for each sheep or bovine.

A very different damage relationship can occur where the season is wetter than usual and there is more grass, and where the stocking rate is low. Here, there is no conflict between rabbits and domestic stock, up to a threshold level, beyond which rabbits do cause a reduction in production.

In this case, the idea of the two extreme damage relationships provided a framework for thinking about the factors affecting rabbit damage, and provided a basis for making rough estimates in particular cases. The decision on the best form of rabbit control, if any, will obviously include other considerations, particularly the cost and effectiveness of control and the rate of population increase.

The form and dimensions of pest damage relationships are also of importance in determining economic thresholds, which we consider later in this chapter.

DETERMINING AVAILABLE OPTIONS, COSTS AND EFFECTIVENESS

Decision Trees

The range of options that can be adopted in managing pest problems can be portrayed in the form of a decision tree; an example is given in Fig. 3.5 for control of a crop disease. Each branch of the tree represents a particular pest management strategy, involving the combination of planting times, varieties and pesticide. There are two main points to make regarding decision trees.

First, a distinction must be made between decision trees that set out the options

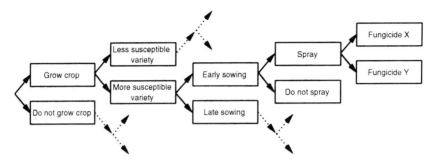

Fig. 3.5. A decision tree for pest management.

which are technically possible, and those that are practically feasible for a particular farmer to use. The latter may have considerably less branches than the former, due to various constraints, such as lack of knowledge about a particular option, appropriate equipment for pesticide application, or simply that there is a shortage of cash to pay for a resistant variety or pesticide. By making the distinction between these two types of decision tree, attention can be focused on the constraints limiting the adoption of some options and the ways in which these constraints might be reduced, thus increasing the range of pest management strategies available.

The second point to note concerning the decision tree shown in Fig. 3.5 is that the nodes of the tree represent decision points over time; in the previous chapter, the decision profile for wheat bulb fly (Fig. 2.11) illustrated this point very explicitly. Thus, the general problem of pest management is that the number of options diminishes, as the season progresses, until the last node – fungicide use (Fig. 3.5). At the same time, information on the level of pest attack usually increases as the season progresses, putting the decision maker in the dilemma of having increased information but reduced options.

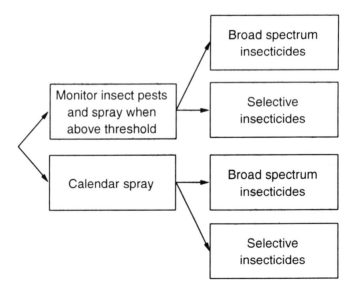

Fig. 3.6. A simplified decision tree for insect pest management in apples.

Thus, used as descriptive tools, decision trees can be of value in several ways. For instance, they can help to:

- Identify constraints that prevent certain options being used in practice.
- Determine the time at which pest management decisions need to be made and, therefore, the time at which forecast or monitored information needs to be available.
- Ensure that all possible combinations of strategies are being considered by research scientists in investigating the performance of control options.

To illustrate this last point, a decision tree (Fig. 3.6) constructed during a workshop on apple pest control (Barlow *et al.*, 1979) identified all the possible options that could be investigated, and particularly those, such as monitoring and spraying with broad spectrum chemicals, which were not being investigated at that time. Thus, by forcing us to consider all the possible options, decision trees provide a valuable framework for deciding which options and strategies need further investigation.

Problem Trees

Another form of tree diagram can be used to explicitly describe the problem and to help set objectives for further action. LaGra (1990) described the problem tree as a way of visualizing the causes and effects of a core problem which has come to the attention of a decision maker.

Figure 3.7 illustrates a problem tree related to fruit flies in the Indian Ocean. In several countries the core problem was seen by local agricultural officials as low local fruit production and they intuitively saw fruit fly control as a way of increasing production. The problem tree (only partially completed in Fig. 3.7), helps to show the nature of the problem more systematically. At each level the problems shown can be turned into objectives. At each level going down the tree the problems and objectives become more specific. This provides a checklist for identifying options to meet the objectives, and can be used to help establish priorities for different options. At the most specific level in Fig. 3.7 some options become clear immediately: provide correct equipment; prune trees; provide training; make baits available. Further questions arise about how to do each of these, but these can also be answered systematically as a result of the structure provided by the problem tree. Looking at other branches of the problem at the same level shows the wider dimensions of the problem which must also be solved. For instance, the relative importance of the wholesaling and investment problems must be determined, and they may also need to be solved to achieve a significant improvement in local fruit production.

Brainstorming Techniques

Brainstorming is a useful way of tapping the imagination and intuition of crop protection specialists when trying to come up with new options for pest management. However, brainstorming needs to be kept directed at the whole target, otherwise there is a danger of concentrating on only a small portion of the problem and totally missing some promising areas.

One way in which brainstorming can be carried out is by using pest life cycle diagrams, in which each target stage or process is identified and brainstorming sessions

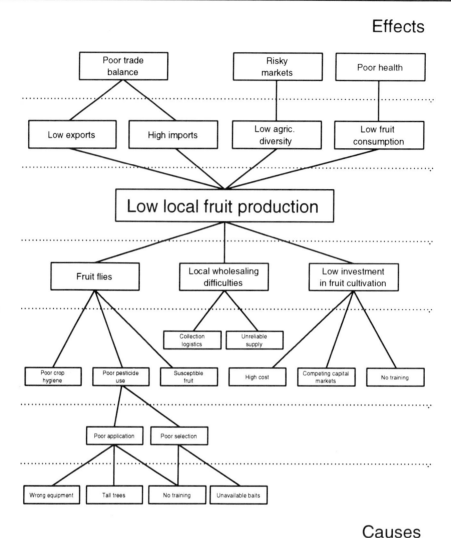

Fig. 3.7. A partially completed problem tree for the core problem of local fruit production in Indian Ocean states. Parts of the fruit fly branch are more completely shown than other branches.

by scientists and extension agents are focused successively on each target. This ensures that the whole range of alternatives is considered. Initially during brainstorming all possible control ideas should be listed for each target. It is important to suspend judgement on these ideas, and get as many ideas as possible (good or bad) written down. Ideas that seem impractical at first may trigger more practical suggestions later. Eventually these ideas can be assessed in terms of such factors as feasibility, cost, effectiveness and research effort needed for development, to concentrate attention on the best options. This can be organized systematically in a matrix, as shown in Table 3.2 below.

To illustrate the value of brainstorming, let us look at a particular example, in this case, a mosquito vector. As shown in Fig. 3.8 the simple life cycle diagram for this pest has been broken up into targets that are qualitatively different from each other, not just developmental stages

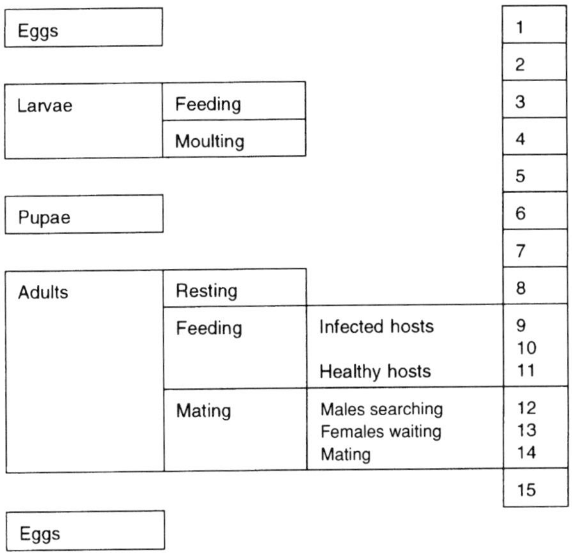

Fig. 3.8. A simple life cycle diagram for a mosquito vector showing qualitatively different target stages and processes, numbered for convenience for brainstorming.

and moults. For instance, larval instars are not (in this case) qualitatively different in terms of control since they are in the same location and feed in the same way. On the other hand, male and female adults may have different behaviour during mating, and so are qualitatively different. They, therefore, may have different control options.

The stages and processes for control options are listed in Table 3.1, which forms the framework for a brainstorming session.

To give some idea of the output that one might expect, Table 3.1 includes part of the list of options generated during a brainstorming session on this topic by participants on a course, who, incidentally, were not mosquito experts. Over 60 suggestions were given in this case and there was considerable discussion of how some of the more immediately impractical ideas could be implemented with some further research and redesign.

Problems can be organized in many ways as a basis for brainstorming. LaGra (1990) listed the individual steps in a production process (commodity production, processing and distribution) and then systematically considered how each of these steps could be improved to give a better overall system. The problem could also be examined by dividing it into time steps (for instance crop growth stages), spatially (for instance, levels of the crop canopy) or any other logical divisions. The aim is to break the problem down into smaller components and to then brainstorm on each of these in turn.

Feasibility Tables

Having produced a list of possible options, as the result of a brainstorming exercise, the next stage is to consider each of these options in a more practical way. This can be done by constructing a feasibility table, where the options are listed vertically and the various criteria that might be used to assess feasibility are listed across the top of the table. Such factors as the likely cost

Table 3.1. Target stages and processes for control (with some examples of options given during a brainstorming session, some of which would not actually be practical).

	Stage or process	Options for control
1	Eggs	Detergent Ovicide Oil on water surface Drainage Fish and other predators Infertile eggs (sterile adults) etc. . . .
2	Eggs hatching	Drainage Flooding Chitin synthesis inhibitor etc. . . .
3	Larvae	
4	Larvae moulting	
5	Pupation	
6	Pupae	
7	Adult emergence	
8	Adults resting	
9	Adults feeding - infected hosts	
10	Adults transferring from infected to healthy hosts	
11	Adults feeding - healthy hosts	
12	Males searching for mates	
13	Females waiting for mates	
14	Mating	
15	Oviposition	

and effectiveness of the measure, how it fits into the farming system, any machinery required, and any environmental or safety constraints, will be included in this list.

To illustrate, a feasibility table for the control of wild ducks that cause damage to rice in Australia is shown in Table 3.2 (Bomford, 1988). This table was used to provide a very preliminary assessment of two groups of options, those which need to be taken before damage occurs (Options 1–7) and those implemented after damage has started (Options 8–14). Six of the 14 options did not meet the feasibility/acceptability criteria, and there was considerable

Table 3.2. A feasibility table for wild duck control in Australia (after Bomford, 1988).

	Feasibility/Acceptability Criteria					
Control options	Technically possible	Practical with farmer resources	Economically desirable	Environmental acceptability	Political acceptability	Social acceptability
1. Grow another crop	yes	no				
2. Grow decoy crop	yes	yes	?	yes	yes	yes
3. Predators and disease	no					
4. Sowing date	yes	yes	?	yes	yes	yes
5. Sowing technique	yes	yes	?	yes	yes	yes
6. Field modifications	yes	yes	?	yes	yes	yes
7. Drain or clear daytime refuges	yes	no				
8. Shoot	yes	yes	?	yes	?	yes
9. Prevent access, netting	yes	yes	no			
10. Decoy birds or free feeding	yes	yes	?	yes	yes	yes
11. Repellants	yes	no				
12. Deterrents	yes	yes	?	yes	yes	yes
13. Poisons	yes	yes	?	no		
14. Resowing or transplanting seedlings	yes	yes	?	yes	yes	yes

uncertainty about how effective the remaining eight options were likely to be. Further investigation was carried out on deterrents, reported to be successful in other areas. Twenty-nine were considered, of which 23 were rejected on grounds of feasibility/acceptability. Research into the cost-effectiveness of the remaining six deterrent options was recommended.

As indicated above, to decide which of the possible and potentially feasible pest management options would be worth exploring further, additional information will often be required. The question is, what type of information, involving minimum effort to acquire, will be sufficient to make this assessment?

During a workshop in Kenya (Prinsley, 1987), the systates weevil, a pest of beans, was found to be of concern to farmers in Embu District. While some reservations were raised during the workshop concerning the importance of this pest, a brainstorming session was carried out to determine what control measures might be worth investigating if further action was to be taken. Possible control options, ranked according to whether they were thought 'questionable' or 'worth following up', are shown in Figure 3.9. Each of these options was then considered in terms of the information that would be required for each control method to be assessed in more detail.

		A	B	C	D	E	F	G	H	I	J	K	L
Adults													
Insecticide	2	■	■	■	■	■			■				
Planting time	1		■		■	■	■						
Tolerant varieties	1				■	■							
Barriers	1		■					■					
Cropping pattern	2	■	■		■	■		■					
Destroy alt. hosts	2		■		■			■	■				
Natural enemies	1								■				
Traps/baits	2	■	■		■			■		■			
Rotation	2	■	■		■	■	■	■	■				
Eggs													
Natural enemies	1												
Ovicidal action	1			■					■		■		
Cultural practices	2				■				■		■		
Larvae													
Cultivation	2				■	■		■				■	
Larvicide	1			■					■			■	
Natural enemies	1								■			■	
Rotation	2	■	■		■	■	■	■	■		■	■	
Cropping pattern	2				■	■		■	■			■	
Destroy alt. hosts	2							■				■	
Tolerant varieties	1				■	■						■	
Pupae													
Natural enemies	1								■				■
Cultivation	2				■		■						■
Rotation	2				■		■		■				■

Key:
1		Questionable value
2		Worth following up
A		Diurnal behaviour of adults
B		Immigration and mobility
C		Chemical control efficacy
D		Phenology
E		Damage relationship
F		Diapause
G		Host range and preference
H		Natural enemies biology/ecology
I		Attractants
J		Oviposition behaviour
K		Larval behaviour
L		Pupal site

Fig. 3.9. Information requirements for potential control options against the systates weevil in Kenya (from Prinsley, 1987).

While this information is grouped under topic headings, it should be noted that the information requirement in each column, for different options, is not necessarily the same. For instance, information on natural enemies required for the natural enemy options will be different for each life stage and will need to be more detailed than information on natural enemies required for assessing the option of crop rotations.

ASSESSING OUTCOMES IN TERMS OF OBJECTIVES

In coming to a decision on a pest control action, the decision maker will attempt to assess the outcome associated with that action and then evaluate it in terms of his or her objectives (Fig. 3.1). In making recommendations, we can attempt to carry out this assessment on the basis of objective information.

Decision-Making Models

Conventional economics approaches resource allocation problems by assuming a deterministic, or at least a perfectly predictable, world. Under these circumstances an action leads to a particular outcome, resulting in a precise relationship between cost and benefit (Fig. 3.10).

Efficiency is a common economic criterion used in deciding on the best action to take, and the production function is the appropriate tool for determining the optimal level of action. An example is given in Fig. 3.11, in this case, for the optimum level of insecticide sprays against cotton pests.

Clearly, this presentation of the problem is very simple on a number of counts but particularly because it ignores the fact that, in practice, decisions are taken in a probabilistic

Fig. 3.10. The conventional deterministic economic model.

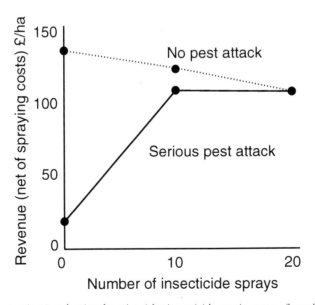

Fig. 3.11. Crop protection 'production functions' for insecticide use in cotton (based on data from Matthews and Tunstall, 1968).

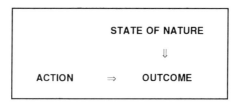

Fig. 3.12. Decision making under uncertainty.

world. This is especially true for pest management, where the level of attack may vary from year to year. Thus, when a particular action is chosen, the outcome that results is one of a range of possible outcomes, depending upon the particular state of nature that occurs (Fig. 3.12). In Fig. 3.11, for example, we assumed that the level of pest attack, or state of nature, was high. However, the dotted line in Fig. 3.11 indicates the revenue associated with different levels of spray when no pest damage occurs. In this case, profit is clearly maximized when no sprays are applied.

Economic Thresholds

Where the state of nature, in this case the level of pest attack, can be monitored in some way before a decision has to be made, the decision maker can choose that action most appropriate to the monitored level of attack. This is the basis for the economic threshold approach, and we can express the decision rule in economic terms, as follows: 'When the level of attack is such that the benefit resulting from treatment is greater than the cost of treatment, then treatment should be applied.'
     This can be expressed in more formal terms, as follows:

$$\text{The benefit of control} = \theta PDK$$

where θ = the level of pest attack (such as the number of insects per plant, the number of weeds per m^2, or the % infected plants); P = the price of the crop or livestock product, expressed in $ per tonne; D = the damage coefficient, expressed as the amount of loss per hectare or the reduction in price caused per unit of pest attack; K = the control coefficient, or the proportionate reduction in pest attack associated with a particular control action.

At the economic threshold level of attack, the benefit of a particular control action (as defined above) just equals the cost of this control action (C), expressed in $ per hectare. That is, where:

$$\theta^* PDK = C$$

Solving for θ^*, the economic threshold level of attack:

$$\theta^* = C/(PDK)$$

While this 'break-even', economic threshold, as defined above, can be determined for many pest problems having a linear damage relationship (Fig. 3.3A), for situations where a threshold relationship occurs, the economic threshold has to be redefined as:

$$\theta^* = T + [C/(PDK)]$$

where T is the damage threshold, that is, the maximum level of pest attack below which losses do not occur (see Fig. 3.3B).

These economic threshold formulae can be useful in obtaining a rough idea of the influence of the different variables on the 'break-even' level of attack for particular control actions. However, for practical decision making, an 'action threshold' needs to be determined. The economic threshold will provide some useful baseline information but to determine an appropriate action threshold, as a practical decision rule, a number of additional factors will also need to be taken into account, including:

- Empirical, 'trial and error' experience.
- The dynamics of the pest population, particularly for endogenous pests, that can build up over time. This raises questions concerning the optimal level and timing of control rather than the much simpler 'break-even' criterion of the economic threshold.
- The risk attitude of the farmer, which will influence the 'margin for error' that will need to be included to make the action threshold acceptable.

As stated above, the economic or action threshold may be an appropriate decision rule where there is available information on the level of pest attack. However, in many if not most situations, knowledge of the state of nature (pest attack) is not available at the time a decision has to be made. For such situations, the pay-off matrix is a more appropriate means of representing the decision problem (Norton, 1976; Reichelderfer *et al.*, 1985).

Pay-off Matrices

The 'pay-off' for each combination of action and state of nature can be shown in a matrix (Fig. 3.13). The outcome associated with each pest control action depends on the level of pest attack that occurs. The initial problem is to assess the outcome of each combination of action and level of pest attack.

The top row, associated with doing nothing, is the damage relationship; in other words, the amount of damage associated with different levels of pest attack. In the remaining cells in the matrix, the outcomes also include the cost of the control action and its effectiveness in reducing pest attack and damage.

Outcomes can be expressed in a variety of ways, whichever is relevant to the decision maker's objectives. A common method is to express the outcomes in terms of net

	State of nature (i.e., level of pest attack)		
	Low	Medium	High
Do nothing	Outcome (L, 0)	Outcome (M, 0)	Outcome (H, 0)
Strategy 1	Outcome (L, 1)	...	...
Strategy 2	...	...	...
Strategy 3	...	...	...

Fig. 3.13. A general pay-off matrix for pest control.

revenue, that is, total benefit minus the cost of any actions, for each state of nature and then to determine the expected monetary value of each action across the entire range of states of nature. This involves multiplying the net revenue for each cell in an action row by the probability of that outcome occurring, which is the same as the probability of the relevant state of nature occurring.

Choosing the option with the highest expected monetary value assumes that the decision maker is risk-neutral. That is, he or she is unconcerned with variations in outcomes from year to year. In practice, most decision makers are likely to have a more risk-averse attitude than this and they may prefer actions that still give an acceptable outcome even when medium to high levels of pest attack occur, even though these may not give the highest overall returns. Where the decision maker is extremely risk-averse, the action chosen will be that which gives the highest outcome when the worst level of pest attack occurs. This is the best outcome in the high pest attack column of Fig. 3.13.

In some cases, it is necessary to make assessments of strategies without solid data on the likely outcomes associated with the various options. One way of resolving such a problem is to rely on expert guesses. This can be refined considerably by using a pay-off matrix to focus expert opinion on the problem and to help resolve any conflicting views.

Figure 3.14 illustrates a pay-off matrix to assess four strategies to deal with a cotton bollworm problem in Africa (Mumford and van Hamburg, 1985). Bollworm attacks are divided into three categories, as shown across the top of the matrix. Down the left side are the four strategies to be compared, doing nothing, weekly spraying, and using thresholds of 5 larvae/24 plants and 8 larvae/24 plants. A blank matrix like that in Fig. 3.13 can be presented to several 'experts' (researchers, extension agents, pesticide salesmen, experienced farmers, etc.) who are asked to complete it. Each level of attack must be given a probability, and then the outcome of each action for each level of attack must be estimated, in terms of the combined loss from pest damage and costs of control. The resulting matrices from the several 'experts' can then be combined to form an average, consensus pay-off matrix.

In Fig. 3.14, spraying when the threshold value of 5 larvae/24 plants is reached comes out as a better strategy than weekly spraying in all cases, on average (expected) cost, cost under worst conditions (high attack category) and cost under the most likely level of attack (medium category). While there may be no solid data to back up such a strategy, this subjective assessment would indicate it was worth testing

	Level of damage (probability)			
	Low (0.33)	Med (0.44)	High (0.23)	Expected cost (loss) per ha
No spraying	40	264	560	254
Weekly calendar spray	163	179	195	177
Economic threshold 5 larvae/24 plants	29	63	78	54
Economic threshold 8 larvae/24 plants	11	45	112	49

Fig. 3.14. A consensus pay-off matrix (in $/ha) for cotton bollworm control strategies, based on the subjective assessment of several 'experts' (after Mumford and van Hamburg, 1985).

such a threshold, given such a large cost difference ($123/ha) and the robustness of the strategy (that is, the wide range of conditions under which it is the better option). In terms of expected returns, spraying with a threshold of 8 larvae/24 plants was thought to be slightly better than the lower threshold, but it is not as robust. In the worst case, with high attack levels, which were thought to occur 23% of the time, this strategy is not as good as the more conservative threshold of 5 larvae. Given the subjective nature of the pay-off matrix, and risk-aversion on the part of farmers, it would not be appropriate to recommend a threshold of 8 on this analysis alone.

Although the calculation of expected outcomes on the basis of the probability of states of nature occurring can provide information of value for decision making, the main point we wish to make about the decision theory approach is simply the value of viewing the problem in terms of a pay-off matrix. Whether one calculates expected values or not, the structure provided by a pay-off matrix gives a valuable framework for further investigation. We will see in later chapters that various techniques, such as simulation models can be employed to help determine likely outcomes of certain actions under different circumstances, and expert systems can be used to help systematically select the most appropriate actions for particular objectives.

While the state of nature in the pay-off matrix set out in Fig. 3.14 is expressed as the level of pest attack, in practice, there may be other factors, such as rainfall and prices, that vary and so affect outcomes. Thus, the state of nature in a particular season may be composed of pest attack, rainfall and prices. While this complicates the issue, a two dimensional pay-off matrix is still a valid way of structuring the problem.

This is not the case, however, where actions and states of nature are connected in time. Where pests are endogenous, for instance, and complete their life cycle within a paddock or a farm, inadequate control can allow pest build-up from one season to the next. For pests such as weeds, nematodes, soil-borne pathogens, rabbits and rats, that fall within this category, the state of nature in the subsequent season is a function of the action (or inaction) taken this season. Thus, a time dimension needs to be included.

Investment Models

A more appropriate decision model for many endogenous pests is the investment model. This decision model is also appropriate in situations where the type of pest control measure used requires an initial large investment that provides benefits in the future. A good example would be investment in an eradication programme, where a pest that is exogenous in the context of the farm is tackled as an endogenous pest, by approaching the problem at a population, regional or national level.

An example for a regional control programme against fruit flies is shown in Fig. 3.15 (Mumford and Driouchi, 1992). A large investment is required initially, with subsequent surveillance and control costs to deal with reinvasion by the pest. This initial cost profile is shown below the horizontal axis in Fig. 3.15. As a result of the eradication campaign, crop production is increased, providing revenue benefits to producers; since these benefits are greater than the cost, the net benefits are shown above the horizontal axis.

A problem that quickly becomes apparent is how to compare costs and benefits that occur at different times. Clearly, a certain amount of money available now is more valuable than the same amount available at a future date, since the present amount could be

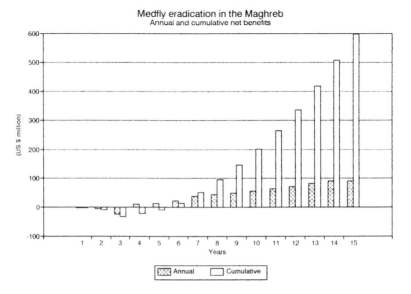

Fig. 3.15. An investment model (net benefit cash profile) for a pest eradication programme (after Mumford and Driouchi, 1992).

invested now and earn additional interest by that date. On this basis, future costs and benefits can be compared by determining their respective present values. This is achieved by discounting future costs and benefits using the formula:

$$V_0 = V_t/(1+i)^t$$

in which V_t is the value at a future time t, V_0 is the present value, i is the discount rate and t is the time in years in the future.

Since the rate of interest used in public projects is always a subject for discussion (Pearce, 1991), it is worth considering the effect that different rates of interest might have on the benefit : cost ratio of a particular scheme. To illustrate, consider again the proposed scheme for regional control of fruit flies in North Africa. A combination of bait sprays and sterile male release could be used to achieve eradication in the Maghreb, providing benefits in terms of increased fruit production and revenue. If we now calculate the estimated benefit:cost ratio of an eradication scheme, initially using a zero discount rate to determine present values, we find that the ratio is 3.01 : 1. That is, for every $1 invested, a benefit of $3.01 is obtained.

Let us now consider a more realistic situation, where a discount rate of 8% is used. In this case, the benefit : cost ratio is lower (2.52), since much of the benefit occurs in the future when its value is less. However, the discount rate is not the only factor that we may wish to vary in assessing the possible outcomes of the fruit fly eradication scheme. For instance, it is difficult to assess what impact on fruit production and fruit revenue the eradication of fruit fly is likely to have. Thus, we can make various assumptions, for instance on the likely rate of progress of the scheme, and consider the economic consequences of these different assumptions. Table 3.3 shows the benefit : cost ratios associated with a combination of assumptions on discount rates and implementation times. This type of analysis, like the pay-off matrix above, can provide a valuable frame-

Table 3.3. Projected benefit : cost ratios for medfly eradication in the Maghreb for three discount rates assuming implementation over 5 years and 9 years (after Mumford and Driouchi, 1992).

Eradication implementation schedule	Discount rates			
	0.00	0.04	0.08	0.12
9 years	3.01	2.75	2.52	2.31
5 years	3.70	2.87	2.60	2.36

work for assessing the significance of different technical estimates on the costs and benefits of such schemes.

An additional problem with large-scale pest management schemes is that although the benefits of success may be very high, the likelihood of achieving that success is very difficult to determine. However, once a government agency has embarked on such a scheme, particularly an eradication scheme, and made the very visible financial and political investment, it becomes locked-in to that scheme, and although it may become apparent that eradication cannot be achieved, it is politically very difficult to drop the eradication scheme.

In New South Wales, Australia, despite a campaign lasting over 70 years, cattle tick has not been eradicated from the state, and government costs to support the scheme have been in the region of A$10 million per year in recent years. Although cattle producers in the area benefit considerably since they do not have to pay for tick control, whether this is the most efficient way to control ticks is difficult to assess. Another factor, of course, is whether the cost of control should be borne by the taxpayer.

Another example of the use of cost–benefit analysis in large-scale pest management programmes is provided by a study undertaken in the USA by Carlson *et al.* (1989). They investigated the benefits of eradicating the cotton boll weevil from a large area in the southeastern USA, showing significant benefits to cotton growers in the area, even with substantial costs associated with maintaining a buffer zone between the eradication zone and adjacent infested areas. However, because of the inelastic demand for cotton, it was estimated that the increased supply of the crop would result in lower prices. The net result would be a considerable loss to many cotton growers in areas of the USA that did not have a weevil problem. However, the producers in the eradication zone would be better off, due to lower control costs and higher yields.

The investment model can also be used as a planning tool for determining the type of information required for making an economic assessment of particular pest management options. To illustrate, consider the case of bush clearing to increase the availability of pasture for livestock production (Norton and Walker, 1985). Depending on the particular method and degree of control, a cost for bush clearing will be incurred in a particular year, as in Fig. 3.16. As grass biomass increases and the live weight of stock increases, the benefits of bush clearing, in terms of increased revenue, will be obtained over successive years (cf. Fig. 3.15).

In making recommendations on bush clearing, some estimate of these cost and benefit profiles must be obtained. The question that faces the adviser is how these costs and benefits are likely to vary from property to property. In other words, what ecological and

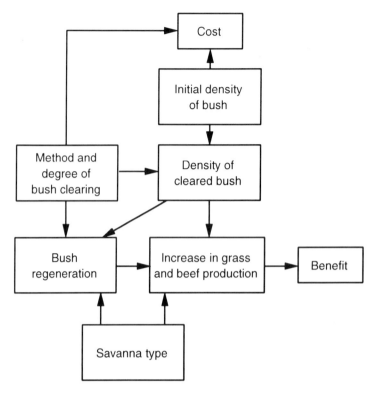

Fig. 3.16. Major factors affecting the costs and benefits of bush clearing.

technical relationships do we need to know to assess the costs and benefits of different bush clearing options? Fig. 3.16 indicates the major processes and states that must be considered in assessing the costs and benefits of bush clearing.

One conclusion that came from this analysis was that a critical factor affecting the benefit profile is the rate of bush regeneration following clearing. Although many field trials of bush clearing strategies have been carried out, there has been little specific study of the critical rate of regeneration. This example illustrates the role that decision analysis and decision models can play in structuring the problem, outlining the important relationships involved, and helping to identify information and knowledge gaps that could form the basis for future research.

Cost Functions

The previous example showed how an investment model, for bush clearing, could be used to identify the major biological and technical components affecting pest management outcomes. A similar approach can be employed using more general cost functions. The example used to illustrate this approach concerns a quarantine pest problem.

In Australia, Foot and Mouth Disease (FMD) has not yet entered the country. If it became established, the losses to the Australian cattle industry could be extremely high. A particular difficulty is that there are feral pigs in Australia which could carry

FMD and provide a reservoir that would be difficult to deal with if the disease became established.

To provide a structure for looking at this problem, a general cost function approach was adopted (Pech *et al.*, 1988). Initially, the question of quarantine effort is ignored. A situation is assumed in which, despite any quarantine effort, FMD is introduced into Australia and into the feral pig population. With this scenario, the question is – What level of monitoring of pig populations should be carried out to detect FMD and so initiate an infected pig eradication programme? The relationship we are interested in is that between the cost of monitoring and the cost of eradication (Fig. 3.17A).

The shape of the relationship in Fig 3.17A will depend on a number of factors, particularly two other relationships:

- The effect of monitoring cost (that is, monitoring effort) on the time taken to detection; the higher the monitoring effort (and cost), the shorter the expected time to detect FMD (see Fig. 3.17B).
- The effect of time to detection on the cost of eradicating FMD in pigs; as time to detection increases, the cost of eradication is likely to increase exponentially (as shown in Fig. 3.14C).

The form of both of these relationships will, in turn, depend on further components. For instance, the form of Fig. 3.17C will be particularly affected by the epidemiological factors affecting the rate of spread of FMD (Fig. 3.17D), and so on. Thus, the initial

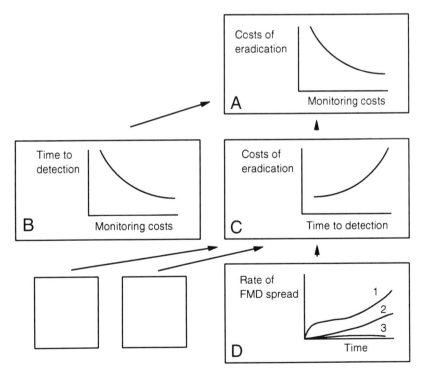

Fig. 3.17. Component cost functions for the monitoring and eradication of Foot and Mouth Disease in feral pigs in Australia. (The empty boxes are to indicate other functions inputting into box C.)

cost function, monitoring vs. eradication, provides a framework for structuring thoughts about the various components of the problem, and how they relate to each other. Further developments of this have led to detailed mathematical modelling of the factors affecting rate of spread (Fig. 3.17D), and to the consideration of novel management strategies. For instance, the idea of taking pre-emptive action in areas where the potential rate of spread of FMD is likely to be high was derived from this analysis.

Rule-based Advisory Matrix

Expert systems are computer programs that can help to analyse, develop, and convey recommendations on pest control. They are discussed in detail in Chapter 11. In the process of constructing an expert system, or simply in creating a logical structure for developing recommendations, a rule-based advisory matrix, such as the simple one shown in Fig. 3.18, can provide a basis for identifying and pulling together relevant information. In this particular example, involving a stored grain pest (*Sitophilus*), the expert was initially asked to list the options available for dealing with a *Sitophilus* problem. These options are shown as the bottom row of Fig. 3.18. The expert was then asked what factors need to be taken into account to determine which is the best option for a particular situation. The 'qualifiers' listed on the left of the matrix are those variables that should be considered in making a recommendation. The expert then considered different situations, defined by different combinations of qualifier values, and represented by the columns in the matrix. The option indicated by the arrow is the recommended one for that particular situation.

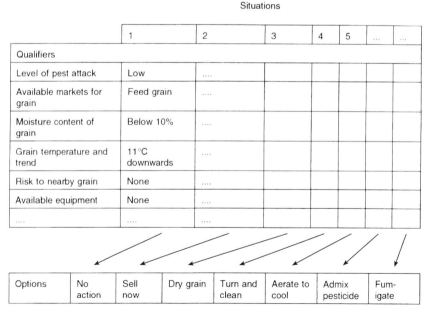

Fig. 3.18. A matrix technique for structuring the knowledge required to make recommendations when a large bulk is infested with grain weevil (Norton, 1987).

Policy Options/Objectives Matrix

Systematic decision analysis involves assessing the performance of different options with respect to the objectives of the decision maker. The example given here concerns the possible response of the Department of Agriculture in Malaysia to an outbreak pest, the brown planthopper on rice (Norton and Heong, 1988). The matrix shown in Fig. 3.19 identifies the major options available to the Department of Agriculture in both the short and medium term, and the possible objectives against which these options are likely to be judged. In this context, scientists can best help policy decision makers by rating each option

Objectives

	More on-farm pest control	More farmer income	More political support	Keep Dept's cost low	Keep pest loss low	Stop pest spread	Less future pest attack

Options

Short							
Warn, advise farmers							
Advise, give credit							
Advise, subsidize pesticide							
Advise, subsidize & guide spraying							
Treat, charge farmers							
Treat at Dept's cost							
Medium							
Intensive surveillance							
Area cultural and bio-control							
Farmer training							

Fig. 3.19. Possible policy options and objectives for the Malaysian Department of Agriculture's involvement in the control of the rice brown planthopper (after Norton and Heong, 1988).

according to each objective. These decision makers can then make their choice on the basis of this explicit assessment, trading off one objective against another.

One purpose of this matrix is to provide the entomologist with a better appreciation of the decision-making context. If this type of analysis had been carried out for the leaf cutting ant project, described in Chapter 1, the criteria of political acceptability and desirability may have been identified for aerial baiting, and the project modified accordingly.

Discrimination Analysis

Farmers' choices of control measures for pests are determined by a number of factors. In a workshop in the Philippines (IRRI, 1992) that focused on control of the golden snail in

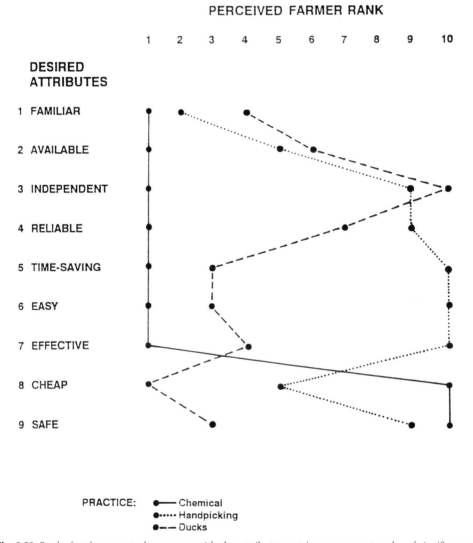

Fig. 3.20. Ranks for three control measures with the attributes put in an apparent order of significance (IRRI, 1992).

rice, a range of attributes for ten snail control measures were considered to be important. These attributes were determined in two ways: from the opinions of workshop participants (scientists and extension officers), and from comments made by farmers in video-taped interviews, prepared in advance of the workshop. Two groups of participants ranked each control method according to each of the attributes. One group ranked the controls on their own expert opinions, while the second group inferred a rank on the basis of views given by farmers in the video interviews. In the case of the attribute 'reliability', for example, each of the ten control measures was put in order, the best being given a value of 1, second best 2, etc. From these rankings two sets of profiles comparing the controls were presented, one for experts' opinion and the other based on perceptions of farmers' rankings. An average ranking, assuming equal weighting for each attribute, was calculated for each control measure. Chemicals came out very favourably in both cases.

In reality, however, farmers would not give equal weight to all attributes. To try to put attributes into the order of importance given to them by farmers, two widely used practices (chemicals and handpicking) and one infrequently used practice (duck pasturing) were compared (Fig. 3.20). Since chemicals are most attractive to farmers, the factors for which chemicals had favourable values were listed first. When several factors had equal values, the remaining factors were ordered by their attributes for hand-picking, because this was practised more than duck pasturing. Finally, if several attributes had equal values for handpicking, the remainder were ordered according to the rank for duck pasturing.

The profile shown in Fig. 3.20 was produced using this method, with the factors listed in order of significance. This is a very rough method of ranking the attributes; a more statistically based sample could be used, and it may be more appropriate to determine an absolute value (cost, effectiveness, etc.) for each attribute rather than a relative ranking. However, this method still gives a good fit to the picture presented by farmers in the surveys, and presents a number of challenges to the introduction of IPM for snail control. Extension must try to make new control 'familiar and simple'; research must produce practical ideas that can be made readily 'available'; and the need for a group response to the community-wide problem of snails must be dealt with by policies that overcome the desire for 'independent' solutions. The lack of importance apparently given to safety by farmers will also require special action by policy makers.

REFERENCES

Barlow, N.D., Norton, G.A. and Conway, G.R. (1979) A systems analysis approach to orchard pest management. Mimeographed report, Environmental Management Unit, Department of Zoology and Applied Entomology, Imperial College, London.

Bomford, M. (1988) Effect of wild ducks on rice production. In: Norton, G.A. and Pech, R.P. (eds) (1988) *Vertebrate Pest Management in Australia.* CSIRO – Division of Wildlife and Ecology, Canberra, pp. 53–57.

Carlson, G.A., Sappie, G. and Hammig, M. (1989) Economic returns to boll weevil eradication. USDA, ERS, Washington, DC.

IRRI (1992) *Report of a Workshop on the Management of Golden Snail in the Philippines.* International Rice Research Institute, Los Banos, Philippines.

LaGra, J. (1990) *A Commodity Systems Assessment Methodology for Problem and Project Identification.* Postharvest Institute for Perishables, College of Agriculture, University of Idaho, Moscow, Idaho, USA.

Matthews, G.A. and Tunstall, J.P. (1968) Scouting for pests and the timing of spray applications. *Cotton Growers' Review* 45, 115–127.

Mumford, J.D. and Driouchi, A. (1992) *Economic Evaluation of Medfly Control in the Maghreb using Sterile Insect Technique. AE/3/1992. International Atomic Energy Agency, Vienna.*

Mumford, J.D. and van Hamburg, H. (1985) *A Descriptive Analysis of Cotton Pest Management in South Africa.* Plant Protection Research Institute, Pretoria.

Norton, G.A. (1976) Analysis of decision making in crop protection. *Agroecosystems* 3, 27–44.

Norton, G.A. (1987) Developments in expert systems for pest management at Imperial College, U.K. *Review of Marketing and Agricultural Economics* 55, 167–170.

Norton, G.A. and Heong, K.L. (1988) An approach to improving pest management: rice in Malaysia. *Crop Protection* 7, 84–90.

Norton, G.A. and Pech, R.P. (eds) (1988) *Vertebrate Pest Management in Australia.* CSIRO – Division of Wildlife and Ecology, Canberra.

Norton, G.A. and Walker, B.H. (1985) A decision analysis approach to savanna management. *Journal of Environmental Management* 21, 15–31.

Pearce, D.W. (1991) *Cost–Benefit Analysis.* Macmillan, London.

Pech, R.P., Norton, G.A., Hone, J. and McIlroy, J.C. (1988) Exotic disease in feral pigs. In: Norton, G.A. and Pech, R.P. (eds) *Vertebrate Pest Management in Australia.* CSIRO – Division of Wildlife and Ecology, Canberra, pp. 58–66.

Prinsley, R.T. (ed.) (1987) *Integrated Crop Protection for Small Scale Farms in East Africa.* Commonwealth Science Council, London.

Reichelderfer, K.H., Carlson, G.A. and Norton, G.A. (1985) *Economic Guidelines for Crop Pest Control.* Food and Agriculture Organisation, Rome, Plant Production and Protection Paper Number 58.

4 Workshop Techniques

G.A. Norton[1] and J.D. Mumford[2]

[1]Cooperative Research Centre for Tropical Pest Management, University of Queensland, Brisbane, QLD 4072, Australia: [2]Department of Biology, Imperial College at Silwood Park, Ascot SL5 7PY, UK

Introduction

In the previous two chapters it was suggested that the system and decision analysis techniques described are often best employed in the context of inter-disciplinary workshops. In this chapter, we address questions concerned with the planning and management of such workshops, particularly those involving research and extension scientists, and possibly farmers and policy makers, all concerned with the same pest problems. The purpose of the chapter is to put forward ideas that could be useful in designing a workshop, to provide a check-list of things that need to be considered in the logistics and detailed planning of the workshop, and to give some idea of the type of outcome that can be expected from the workshop. There are three sections:

- An outline for pest management workshops.
- Activities to undertake before the workshop.
- Recommendations of a workshop: Rice pest management in Malaysia.

An Outline for Pest Management Workshops

In this section we concentrate on the overall strategy of the workshop, and the factors that are likely to be included in setting objectives, planning (including timetabling), and determining the output of the workshop. Some further ideas on workshop management have been presented by LaGra (1990).

Workshop Strategy

Although there may be a number of reasons for conducting a pest management workshop (see objectives below), the major aim is always likely to include the identification of the best means of improving pest management. To do this, we firmly believe that, initially, an

Table 4.1. A general workshop format.

PHASE 1 (Problem definition: specification)
- What are the key factors and processes determining the overall problem?
- How are these factors and processes changing over time, and what is likely to happen in the future?
- What are the current controls practised, and why?
- What are the key biological relationships that determine the benefits and risks of management strategies?

PHASE 2 (The search for solutions)
- What are the possible options for improved management strategies?
- What constraints and objectives affect the choice of these management options?
- How can options be modified to be more appropriate to on-farm cropping systems?
- What key knowledge and information do we need to obtain and disseminate in order to improve management practices?

PHASE 3 (Recommendations)
- Policy recommendations
- Applied research priorities
- Advisory/implementation priorities
- Management recommendations

---

objective analysis of the problem has to be undertaken. This provides a sound basis for identifying the major factors that need to be investigated further if practical improvements are to be achieved, and a realistic context in which the search for potential solutions can be undertaken. Thus, at the end of the workshop, a clear statement of the problem can be made, the most likely means of improving pest management identified, and recommendations produced on the research, extension, training and policy actions required to increase the chances of this improvement being achieved. Following this rationale, the structure of the workshops usually follows the format given in Table 4.1.

Since one of the purposes of a workshop is to derive recommendations from a systematic and comprehensive analysis of the problem, it is important that the analysis and findings obtained from Phases 1 and 2 are well digested before attempting to draft recommendations (Phase 3). This could be achieved by setting aside a period for participants to reflect on what has been revealed. Alternatively, and probably a much more feasible option given the pressures at this stage in the workshop, is for two or three participants/resource persons to be designated the job of summarizing the major findings of the workshop just before Phase 3 commences. This can then be discussed to achieve an agreed list, which will form the basis for determining recommendations.

Let us now look in more detail at the objectives of the workshop, what is involved, a typical timetable, and what the output of the workshop is likely to be.

Workshop Objectives

The main objectives of a workshop will obviously depend on the particular situation. Some are largely concerned with providing training to research and extension scientists in the value and use of decision tools, that can be of subsequent help to them in their own work. In this case, the prime purpose of the workshop is to introduce a range of techniques and give participants experience in their use by focusing on a particular pest management problem. Other workshops are more concerned with determining what needs to be done in searching for ways to reduce immediate pest management problems. Whatever the main emphasis of the workshop, however, most are likely to have some elements of the following objectives:

- Provide an explicit and structured definition of the pest management problem.
- Identify and evaluate current and potential options/constraints.
- Develop an independent assessment of priorities.
- Provide consensus on recommendations for action.

General Plan of the Workshop

Following the three phases of the workshop, described above, the major activities described below are likely to be included in the workshop, in the sequence shown:

- Introduction to workshop philosophy
- Introduction to decision analysis techniques
    Planning and design
    Implementation
    Sustainable management
- Presentation of reports (prepared in advance) on problem status
    Relevant institutions involved in the problem
    Current pest management activities and their rationale
    Degree of pest management success/failure
    Inputs needed for sustainable management
- Workshop sessions on analysis of current pest management
    Historical development of the pest management problem
    Case studies
    Comparisons with other areas
- Workshop sessions on options/constraints
    Identification of technical/policy options
    Additional inputs needed to implement options
    Adaptive strategies to keep options open
- Workshop sessions on priorities
    Institutional objectives defined
    Criteria for evaluating options
- Recommendations for action
    Policy strategies
    Technical developments
    Organization/infrastructure developments
    Training needs

To show how this general plan can be put into action, a typical timetable for a workshop is given in Table 4.2.

Table 4.2. A typical workshop timetable.

Day 1	am	• Introduction to participants • Introduction to workshop philosophy and techniques
	pm	• Workshop objectives and timetable • Reports on workshop issues (prepared in advance) • Discussion of reports • Determination of priority topics for field visit and workshop session I
Day 2	am	• Workshop session I (e.g. subgroups on historical profiles, future scenarios, planning field visit interviews, etc.)
	pm	• Presentation of subgroup reports (from session I) • Discussion of presentations • Determination of priority topics for workshop session II
Day 3	am	• Field visit (where feasible) (e.g. subgroups interviewing farmers, extension agents, etc.)
	pm	• Workshop session II (e.g. subgroups on seasonal profiles, decision charts, interaction matrices, etc.)
Day 4	am	• Presentation of subgroup reports (from field visit and session II) • Discussion of presentations • Determination of priority topics for workshop session III
	pm	• Workshop session III (e.g. subgroups on information and training needs, research and implementation priorities, etc.)
Day 5	am	• Presentation of subgroup reports (from session III) • Workshop session IV (subgroups preparing draft workshop results)
	pm	• Final presentation and discussion of results • Agreement on statement of workshop results • Closing remarks

Output of the Workshop

The product of the workshop is likely to include the following major features:

- Independent analysis of the workshop problem.
- Structured information exchange on issues determined by workshop objectives.
- Training in decision analysis for participants.
- Recommendations on key research and implementation priorities, as obtained from objective problem analysis.
- Written summary of workshop results (an example is given at the end of this chapter).

ACTIVITIES TO UNDERTAKE BEFORE THE WORKSHOP

The success of a workshop depends not only on the management of the workshop itself but also on the degree of effort made prior to the workshop. This includes the choice of location, data gathering, information supplied to participants, and the equipment needed for the workshop.

Choice of Location

The location at which a workshop is held will depend on a number of factors. Some of the more important factors to consider are:

- If the workshop is to focus on one particular pest problem, there should be easy access to farmers and local extension officers, who are experiencing the problem.
- Facilities which provide both accommodation and the necessary facilities for the workshop are preferable. In particular, this includes a room with overhead and slide projection facilities that will seat, preferably in a circle, around 30 or so people (workshop participants and other 'visitors', such as local extension officers). In addition, two or three small rooms should be available to accommodate group discussion.

Data Gathering

There are a number of topics on which information could be gathered before the workshop, including details on:

- Current agricultural practice, pest problems and pesticide use
- Historical events
- Seasonal events
- Farmer practice

Once collected, this information can be presented to participants at an early stage in the workshop to provide a background to pest problems in the particular locality being studied.

Some guidelines, or check-lists, for the collection of this information are given below.

Background Information on Current Practice

Information on the area of crops, production levels, irrigation and marketing practice, pesticide use levels and other appropriate background information can be obtained from a variety of sources, ranging from government statistical departments, international agencies (such as FAO), non-government organizations and marketing boards to exploratory interviews and group discussions with local extension officers, key farmers, commercial operators and other relevant parties.

Historical Profile

A historical profile of a pest problem consists of a series of graphs, where the horizontal axis covers the historical period to be reviewed up to the present, and the vertical axes represent quantitative or qualitative scales of the major factors related to the pest problem. To complete the profile, estimates of the major trends in these factors need to be recorded. Examples of two historical profiles were provided in Chapter 2.

When drawing up a historical profile a reasonable length of time should be covered (at least 10 years), depending on the data and memories available. The list of topics to be covered in the historical profile might include:

- Policy developments (especially the introduction of new Acts and regulations and subsidies relating to pest management (e.g. pesticide subsidies)).
- Price of the crop(s), including changes in guaranteed prices.
- Crop varieties planted, if possible, expressed as a percentage of the crop area.
- Adoption (percentage area) of other production practices, such as mechanization, direct seeding, etc.
- Changes in the status of pest problems: expressed either as percentage area heavily affected or a rough classification of high, medium and low attack years.
- Adoption (percentage area) of crop protection practices, such as herbicide use, type of sprayers used, type of insecticide, etc.

Seasonal Profile

A seasonal profile charts the major changes and events that occur through the cropping season that may in some way be related to pests and pest management. A seasonal profile is best constructed in collaboration with local extension agents and farmers, using the framework of the profile to extract information from them. The following items are likely to be relevant:

- Annual variation in climatic factors, particularly temperature and rainfall, but also events like typhoon periods, etc., where relevant.
- Annual variation in crop prices, hired labour, etc.
- Cropping practices: such as irrigation times, planting times, cropping sequences, cultivation practices, weeding times and pesticide spraying.
- Farm labour availability through the season.
- Incidence of pests (and different life stages, where relevant) throughout the crop season, as well as the location of pests in the between-crop period.
- The major decision points through the season and the options and information available at each decision node.

Farmer Interviews

There are a number of ways in which interviews can be carried out, including exploratory informal surveys with extension agents and farmers, more formal knowledge, attitude, perception (KAP)-type of interview surveys, focused group interviews and rapid appraisal techniques. Further discussion on interview techniques can be found in Chapter 5.

The choice of technique will depend in part on the time and expertise available. What we would stress is that, whatever the technique used, the following points are important:

- The objectives of the farmer survey are very clearly stated.
- The data required to meet these objectives are precisely specified.
- The survey is carefully designed to collect the specified data.

Information for Participants

When inviting participants to attend the workshop, it is valuable to provide them with background information concerning the overall purpose of the exercise and the likely goals of the particular workshop. This might include one page that gives the overall philosophy of the approach (see Outline section above). In particular, the systematic, interdisciplinary analysis of the pest problem to be carried out in the workshop should be emphasized, which then provides a framework in which to discuss and identify key research, extension and policy priorities.

If appropriate, participants could be invited to prepare background material on the problem in their particular region. In particular, historical and seasonal profiles are likely to be most useful. This will provide some basis for discussing differences between the situation studied in detail in the workshop and that in other regions, and the extent to which the conclusions drawn for the workshop location are applicable to a wider area. This material, from other areas, might best be included in the workshop either as a separate session towards the end of the workshop, or as evening sessions throughout the course of the workshop.

Workshop Equipment

There are a number of items that are essential for the smooth and efficient operation of workshops, including:

- A number of overhead projectors can be used during individual group sessions (say, three or four projectors), as well as for presentations to all workshop participants. Remember to have spare bulbs available!
- One or two slide projectors: with spare bulbs.
- Five or six large white boards (preferably) or blackboards, that are either free standing at a convenient height or which can be propped up conveniently.
- At least one box of 100 transparency sheets and permanent ink pens to go with them. Water-soluble pens, while allowing easy correction, can result in valuable information being lost during the post-workshop period.
- A supply of pads, pencils and pens for participants.
- Computer facilities, including printers, that will provide word processing facilities on site, and which will also allow the demonstration of database systems, expert systems and mapping programmes, if required.

RECOMMENDATIONS OF A WORKSHOP: RICE PEST MANAGEMENT IN MALAYSIA

To give the reader a better idea of the type of results that can be obtained from a well planned and organized workshop, the rest of this chapter presents the report prepared on the basis of a workshop held in Alor Setar, Malaysia, in 1990. The focus of the workshop was on the range of pest problems recently and currently experienced in the Muda Agricultural Development Authority (MADA), the largest irrigated rice scheme in Malaysia, producing 40% of the country's rice output.

Summary

To undertake a systematic and rigorous analysis of pest and pest management problems in the MADA scheme a one week workshop was held at the MADA headquarters, Alor Setar, from 13 to 17 August, 1990. The purpose of the workshop was to use a range of decision tool techniques to investigate the major pest problems in the area and to identify implementation, research and training needs. This report describes the activities and results of the working groups and their subsequent recommendations.

Recent developments affecting pests and pest management in the MADA area are described in the form of an historical profile. This provides a basis for investigating future scenarios and their implications for pests and pest management.

Interviews with case study farmer groups provide some insight into the opportunities and constraints for improved pest management. A comparison of IPC and non-IPC farmer performance, using a pay-off matrix based on expert opinion, provides a further means of exploring these factors.

A seasonal profile of a typical cropping system in MADA allows a more detailed description of the farming system. In this context, pest profiles for brown planthopper (BPH), tungro, *Echinochloa* and rats are used to identify the major components and decision points for control that occur through the season.

Against the background set by these descriptive techniques, workshop participants then went on to develop prototype rule-based advisory systems (or expert systems) for three major pests – BPH, tungro and weeds. An analysis of pest surveillance, its objectives and future opportunities, is also described.

Recommendations

On the basis of these descriptive and prescriptive exercises, workshop participants then produced the following recommendations for further action (further details were included in the report).

1. Undertake a pilot project on the use of decision tools at the farm level

Focusing on eight group farming schemes – four progressive schemes and four schemes with important problems – the aim of this project is:

- To obtain a better appreciation by MADA, MARDI (Malaysian Agricultural Research and Development Institute), DOA (Department of Agriculture) and the farmers themselves of the specific group's problem.
- To identify feasible options for reducing constraints to improved pest management.
- To produce and evaluate farmer-level decision charts for practical day to day use.

2. Increase research on direct seeded rice

Research in the past, and much current research on rice pest management, is concerned with transplanted rice. Since direct seeded rice now accounts for over 90% of the crop area in MADA, there is an urgent need for research to be undertaken on direct seeded rice, particularly on the following topics:

- Pest monitoring techniques **for use by farmers**
- The relationship between agronomic practices and pests
- Resistant varieties
- Natural enemies
- Pesticide application
- Socioeconomic studies

3. *Develop decision tools for rice pest control in Malaysia*

It is recommended that the existing National Integrated Pest Control (IPC) committee initiates an inter-organizational effort to develop computerized expert systems for giving advice on the management of major rice pests, such as tungro, BPH, weeds and rats. This will serve to:

- Identify information needs at the extension and farmer level requiring research and training inputs.
- Provide a means of training extension agents in the management of these different pests.
- Serve as a means of improving communication between research and extension scientists.
- Provide a basis for developing simplified, practical decision aids for farmers.

4. *Develop training programmes for extension agents and farmers*

It is suggested that appropriately designed training courses should be conducted to change attitudes, to improve interaction among individual farmers and between farmers, extension agents and researchers, and to provide technical training.

In designing training courses for farmers, effort must first be devoted to determining what farmers need to know if they are to carry out good management of a particular pest. Then a farmer survey should be conducted to determine the extent to which farmers already have this knowledge, and training should be focused on those areas where the information gap is greatest.

In-service training courses for extension agents should include the application of decision tools, and self-evaluation skills for evaluating the impact of extension activity for on-course correction.

5. *Surveillance workshop*

It is recommended that a special workshop be organized to review the existing rice pest surveillance systems in Malaysia and to explore opportunities for improving their relevance to decisions being made by farmers, extension agents, research scientists and policy makers.

6. Follow-up workshop

It is suggested that a follow-up workshop be conducted within 18 months to 2 years of
this initial workshop, to monitor the progress made on the recommendations above and
to undertake an up-date analysis of the problems at that time.

REFERENCE

LaGra, J. (1990) *A Commodity Systems Assessment Methodology for Problem and Project Identification.*
    Postharvest Institute for Perishables, College of Agriculture, University of Idaho, Moscow,
    Idaho, USA.

5

Survey and Knowledge Acquisition Techniques

J.D. MUMFORD[1] AND G.A. NORTON[2]

[1]*Department of Biology, Imperial College at Silwood Park, Ascot SL5 7PY, UK:*
[2]*Cooperative Research Centre for Tropical Pest Management, University of Queensland, Brisbane, QLD 4072, Australia*

INTRODUCTION

In Chapter 1, constraints associated with the adoption of supervised control by apple growers were described. One implication drawn from this was that if researchers had better appreciated the apple growers' situation, better targeted research and a more acceptable control strategy may have been developed. Part of the problem is that scientists and farmers often have very different approaches to pest management.

In another example, a project which looked at ways of improving the pest control decisions of vegetable growers in the Thames Valley, England (Jusoh and Norton, 1987), revealed differences between the growers' perception of the problem and that of entomologists. An entomological view of the problem is that there is a need for increased monitoring of pest numbers in the crop that will enable growers to utilize action thresholds, which indicate when the grower should spray.

In practice, after talking to growers, it was realized that the complexity of the farming system is such that the grower often does not have the management flexibility required for an action threshold strategy. Indeed, growers view the problem in a very different way. Since sprays can only be applied on days which are not windy, when the soil is not too wet, and when rain is not likely within a day or so after spraying, there may be very few spray-days within a particular month. Thus, when a day is suitable for spraying, the problem is to decide which crops to spray within that limited period of opportunity. Although this is an overstatement of the difference, growers and entomologists certainly do have a very different perspective of the problem.

These examples emphasize the need to understand why farmers manage pests in the way that they do and to try to predict how they will respond to changes in pest management suggested by scientists. Surveys are a useful tool to help gain this understanding and to provide the inputs into the decision model described in Chapter 3. This chapter presents an introduction to survey techniques for pest managers. For those who want a more detailed account of survey techniques various additional references are given throughout the chapter.

WHY CONDUCT SURVEYS?

In pest management, as in any form of management, it is important to know both directly observed facts about a problem and subjective and objective information from people (farmers, managers, etc.) concerned with the problem. Improved pest management is a product that must be sold or delivered in a way that meets the needs of the customer. As in any other business, market research is needed to establish what the customer wants and how it can be provided.

Survey information can be of value in determining the following:

- Identifying the crop production and crop protection measures used in practice.
- Assessing opportunities and constraints for improving crop protection, and assessing the feasibility of novel techniques and practices.
- Assessing the need for information at the farm level.
- Forecasting the decisions that farmers are likely to take in the future.
- Obtaining realistic and relevant criteria for evaluating options.

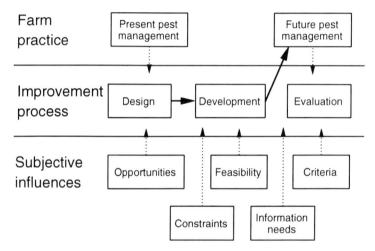

Figure 5.1. Information on practices and subjective influences on the implementation of improved pest management can be of great value in planning. (Technical and ecological influences are not included in this figure.)

As shown in Fig. 5.1, the information provided by surveys can be of value in the design of research and development programmes as well as advisory and training programmes.

Surveys allow a manager to get information on:

- Objectives
- Behaviour
- Attitudes
- Values
- Unpublished facts

This information may come from scientific or management experts, specific potential users of management services, or the general public. Because there are different target groups for surveys, different methods may need to be employed, even within the

		Feedback mechanisms					
		Farmer surveys			'Expert' surveys	Statistical data	Ecology survey
		Semi-structured open question	Focused group interviews	Structured question survey	Agricultural officers, scientists	Pesticide use	Pest status
Objectives	General background information on problem	★★★	★★	★	★★★	★	★
	Determine major constraints to improvement	★★	★★★	★	★	–	–
	Determine information gaps	★	★★	★★★	★	–	–
	Assess feasibility–desirability of technology	★★	★★★	★★★	★★	–	–
	Search for farmers with solutions to problems	★★	★★	★	★	–	–
	Assess the impact of extension–training programmes	★★	★★	★★★	★★	★★	★★★

Figure 5.2. Types of feedback mechanism suitable for particular objectives.

same survey (see Fig. 5.2). The rating of feedback mechanisms in Fig. 5.2 is subjective. Keep in mind that it is usually desirable to use a variety of techniques to ensure getting a good picture of what is happening.

A considerable number of survey techniques have been developed in recent years for conducting farm and rural land use surveys (Dillon and Hardaker, 1980; Chambers *et al.*, 1989). Two main groups of techniques are widely used, collectively known as KAP (Knowledge, Attitude, Practice) (Adhikarya and Possamentier, 1987) and RRA (Rapid Rural Appraisal) (Mascarenhas *et al.*, 1991). KAP surveys provide baseline information on the current situation and give some indication of what people feel about future problems and innovation. RRA provides the same sort of information, but uses some different techniques. It is generally less structured and involves more participation by groups of respondents, rather than individuals. It also uses more graphical techniques, such as asking respondents to draw maps, or to show relationships between ecological and institutional aspects of the problem under study. Some of the techniques that can be used in these group sessions have already been discussed in Chapters 2 and 3. This chapter will concentrate on more formal methods of surveying; some general principles for surveys are set out below.

OUTLINE PROCEDURE FOR CONDUCTING SURVEYS

Ten steps are involved in conducting a successful survey:

- Set clear objectives for the survey
- Establish the population to be sampled
- Consider sample size needed
- Consider survey method
- Design questionnaire
- Conduct pilot survey
- Revise questionnaire
- Conduct main survey
- Code results
- Analyse results

Each of these is discussed in more detail in the following sections.

Objectives

To set the objectives for a survey it is important that those designing it first conduct a short survey on themselves to answer the following questions:

- What particular items of information are wanted?
- Why are they wanted?
- Who will make use of the information?
- How will it be analysed?
- How will it be used?

Some preliminary description of the problem being addressed should also be made, using some of the techniques suggested in Chapter 2, to ensure that the issues to be investigated in the survey are quite clear.

Sample Population

It is important to determine who has the information that is wanted. The population should be delineated precisely by some characteristics, for instance, age, location, interest group and occupation. Identifying the sample population carefully can greatly improve the efficiency of the survey, by keeping costs low and increasing the chances of good responses.

Preliminary screening may be needed to identify the sample, for instance a sample of farmers could be drawn from lists held by local agriculture officers or other officials, but a preliminary question about size or type of farm could narrow down the population to farms of the most appropriate type for a subsequent interview sample.

Sample Size

The sample size depends on the objectives of the survey. If you are just trying to get a feel for the issues and the range of opinions, a small sample may be adequate, but if you want to present quantitative conclusions on opinions or facts, the sample will need to be larger.

The size depends on both the level of precision needed in the results and on such practicalities as the available resources of time, effort and funds. The more precision needed and the greater the expected coefficient of variation in the responses, the larger the sample size must be to achieve them.

Many statisticians and market researchers give advice on sample size (see, for example, Siegel, 1956; Tull and Hawkins, 1976). Different methods for calculating sample size should be used to estimate sample sizes where means are involved (average farm size) and for questions of proportion (percentage of farmers growing maize). In either case you must make an estimate of both the expected response and the variation in response for each question. These calculations should be done for each question in the questionnaire, with the overall survey size determined by the question requiring the largest sample. This makes a pilot survey essential.

The statistics used to determine the sample size required for a particular survey can provide very sobering results. It generally takes a very large survey to achieve meaningful statistical results, and those about to conduct a survey should realize this at the outset. For example, if you expect 40% of respondents to answer a question in a particular way and you will accept a 5% error around that value (35–45%) with 95% confidence then the sample must have at least 400 respondents (completed replies). Postal and media surveys need much larger sample sizes than interview surveys (personal or telephone) because of their large non-response. The statistics give the size of the *returned* sample needed, not the attempted sample size.

An alternative use of these statistical methods is to see the level of precision that is likely to be achieved from a given sample size (determined by logistical limits, for instance).

Survey Methods

The advantages and disadvantages of three types of survey are outlined below. Unstructured techniques, in which questions are not preset, and individuals or groups are guided in a discussion on the subject of interest, can only be used in personal interviews.

Personal interview

- Face to face or by telephone.
- Good for opinion-based surveys where there are complicated issues that need follow-up questions and considerable interpretation.

- Necessary for certain types of populations, for example, pesticide purchasers could be more efficiently approached personally at a pesticide shop, rather than visiting farms at random.

Postal questionnaire

- Cheap way to survey for relatively straightforward opinions and facts.
- Drawbacks include low response rate (generally 30% or less), uncertainty over interpretation of questions, no chance to check validity of answers, potential bias towards people with strong opinions, unsuitable for people with reading/writing difficulties, no fixed address, etc.

Media questionnaire

- Cheap way to survey a large group of people with interest in a specific magazine, newspaper, radio, or television programme.
- Response rate generally low, may need to offer 'prize' as an incentive, bias to people with strong views or desire for prize, plus drawbacks of postal questionnaires.

Questionnaire Design and Types of Questions

The key to successful questionnaire design is pretesting questions on potential respondents. This will quickly show which questions will give sensible responses and which are misunderstood or cause difficulties for the respondent. The following points provide some guidance on things to remember when planning questions.

- Ask questions directly, with as few words and subordinate clauses as possible. Use plain English (or other language, as appropriate), using common local words and familiar terms. Keep it simple!
- Make the language fit the target population. For instance, in a survey of farmers you might ask about pests by their common names or showing them specimens, whereas a sample of entomologists may be asked questions using the scientific names.
- Do not lead the respondent on opinion questions, keep questions neutral.
- Ask positive questions about what respondents do or think rather than questions about what they do not do or think. For example, 'Do you use pesticides?', rather than 'Do you avoid using pesticides?'
- Begin a questionnaire with some simple factual questions to ease the respondent into the subject, questions to which everyone has an easy answer (such as the size of the farm and what crops are grown). This helps to establish a non-threatening rapport with the questioner.
- Do not ask potentially embarrassing questions, such as age, income, occupation (if there is a chance of unemployment), education level, etc. at the start of an interview. If it is essential to get such information, then leave these questions to the end of the interview.
- When asking about potentially embarrassing behaviour, or controversial attitudes that may be contrary to public norms, it is possible to ask people for their opinion on how many of their neighbours have that attitude or behaviour. For example, How often do your neighbours spray market vegetables near harvest? Their own situation may be reflected in their view of other peoples' actions or thoughts.

- Only ask those questions for which you need to know the answer, no more, no less. An exception might be simple 'startup' questions, or questions that do nothing but improve the flow of the questionnaire by avoiding abrupt changes in direction during the survey. Keep the questionnaire as short as possible.
- Some people have difficulty with percentages, rates and fractions, so be aware of this in designing questions. It is often better to ask for two absolute figures rather than a percentage or rate. As an example, 'How many tonnes of fertiliser were applied to the crop?' and 'To how many hectares was it applied?' The rate can then be worked out later.
- To get estimates about subjective values held by respondents it may be better to ask direct questions about expenditure or time devoted to an activity, or costs of alternative activities/options, rather than to ask for an abstract value estimation.
- When asking multiple choice questions respondents tend to remember the first and last in a list of potential answers, which could give biases to the results. If the list of choices is long or the choices are not in a particular pattern, for instance names of insecticides, rather than a good–bad scale, it is a good idea to provide the list on a durable card that the respondent can look at while answering. Don't have too many cards (both you and the respondent may get confused), and keep the wording on the cards very simple so that a respondent can read it all quickly. Consider picture cards if there may be reading or language problems and keep print size large.
- Scale and multiple choice questions are easier to code and analyse than open answer questions, and they are also quicker to record while conducting the survey.

There are four types of questions:

- Fact with wide response range – name, address [open answer]
- Fact with narrow response range – type of housing [multiple choice]
- Opinion with wide response range – what is the most pressing catastrophe the world faces today? [open answer]
- Opinion with narrow response range – attitude on a scale, or preference among a few choices [attitude scale/multiple choice; subjective values]

When designing scale questions make sure that they balance, e.g. very good, good, no effect, bad, very bad. A scale such as the following is NOT well balanced: excellent, very good, somewhat bad, moderately bad. If you have an even number of choices in the scale you will not get a middle response (although this may be desirable if you want to force a commitment one way or the other). Three points on a scale is minimal, four or five is reasonable. If you have a larger number, for instance a 0 to 10 scale, you may find that people are unwilling to use the end points, so it may effectively become a 1 to 9, or 2 to 8, scale.

Pilot Survey

It is essential to pretest a questionnaire design on a small group to ensure that the questions are interpreted properly, to get a feel for the range of answers that are likely to be given, and as practice for the questioner (particularly in interview surveys) to get a good flow of questions. Recognize that the average respondent is not necessarily going to view your questionnaire or the issues with which it deals with the same logic that you have imposed on it by thinking about it for some time.

Revision

Changes will almost inevitably need to be made in the questionnaire design following the pilot survey. Do not make hard and fast plans for the main survey (i.e. printing thousands of questionnaires, etc) until you have had time to consider the results of the pilot.

Main Survey

Keep as much of the survey activity as consistent as possible. For instance, try to use the same interviewers and style of questioning. Unless the survey is looking at changing attitudes over time it is better to do it all over as short a time as possible.

Make additional notes on interview respondents as soon after the interview as possible, and check through the questionnaire form immediately after the interview to make sure all is legible. Do not leave these tasks for several weeks.

Coding

Coding is the process of transferring data from questionnaires to the form in which it will be analysed. Keep coding for computer analysis in mind when designing questions and forms. Scale and multiple choice questions are easier to code than open questions.

Open questions will need to be classified into groups and coded accordingly, except for straight facts like name and address. Even these may need to be coded for ease of analysis, such as people living in a village or outside the village. Be careful how you define groups; for instance are the outskirts of a village in or out of the village in the context of your particular issue?

Analysis

For small surveys analysis can often be done simply by hand, or on a computer spreadsheet. This involves making summary tables and relatively few figures. For larger surveys, computer analysis is essential. The simplest way to analyse survey results is to use computer programs such as SPSS, SAS, Minitab, Genstat, CSS and dBASE.

Much of the analysis will take the form of listing and sorting, and presenting graphs and tables. However, some statistics may be needed to differentiate groups or respondents.

There are three basic types of statistics that can be used, depending on the type of data that has been collected:

- Contingency table tests (Chi square or Fisher's Exact tests) should be used for frequencies of classed data. Classed data is simply information on the frequency of classes of responses which do not have numerical relationships, for instance yes/no, red/blue/green. There is no order or quantitative relationship between the classes.

- Rank tests (such as the Mann-Whitney or Median tests) should be used for ordinal data. Ordinal data is numeric or scaled by rank but without a quantitative relationship. With a very good to very bad scale of 5 points you can say that point 4 is less than point 5, but not that it is 80% as bad as 5. These tests let you decide if respondents are giving answers with different ranks.
- Mean tests (such as the t test) should be used for rational data. Rational data is numeric data which has a quantitative relationship between the various possible values; for instance a field of 10 ha is twice as large as one of 5 ha.

Special Features of Expert Surveys: The Delphi Technique

If you are trying to make a prediction of a future event, for instance the impact of a new form of pest control with which farmers are unfamiliar it is often desirable to test the opinion of 'experts' in the field. In cases in which the expert opinions are used to determine management objectives or to make important decisions it is important to ensure that a consensus opinion is reached. This may be difficult to achieve by sending out a single questionnaire and then trying to evaluate the importance of the various divergent opinions expressed. However, a more valid consensus can be achieved using the Delphi technique, a series of surveys allowing feedback to experts before they are asked to come to a consensus.

Initially, experts are asked to give individual, anonymous responses to questionnaires. The overall results are tabulated and returned to each participant, who is then asked to revise his/her answers in the light of the other respondents' views. This technique allows the experts themselves to evaluate their own confidence in their opinions relative to those of their peers. The process may be repeated one or more times until a satisfactory level of agreement has been reached.

Anonymity is important in the early round(s) of the process to ensure that experts are not put in a position where they cannot back down on their views without losing face.

A Delphi approach can also include a preliminary questionnaire, followed by group meetings with experts, at which they can defend or relinquish their positions after considering the views of the others. This may be practical if there are relatively few experts being surveyed. For example, this approach was used by Mumford and van Hamburg (1985) to develop and test an expert system on cotton pests in South Africa. Five experts were independently asked to give advice about rules for cotton pest control. The rules were compiled on the basis of the individual inputs and then all five met together to discuss them and to develop a consensus view.

A major example of the use of the Delphi technique comes from the USDA (1981), who used it to evaluate the expected benefits of a potential regional eradication campaign against the boll weevil in the USA. A group of 35 experts involved in cotton research, extension and production, along with chemical company representatives and farmers, were questioned in three rounds. After the first two rounds, summaries of responses were given to the experts who were then given a chance to revise their estimates of the benefits, in terms of increased yields and reduced pesticide costs associated with six different potential insect management strategies. The results were then used by policy makers as one of several inputs in deciding on the political question of whether to proceed with the programme.

REFERENCES

Adhikarya, R. and Possamentier, H. (1987) *Motivating Farmers for Action: How Strategic Multi-media Campaigns can Help*. GTZ, Eschborn, Germany.

Chambers, R., Pacey, A. and Thrupp, L.A. (1989) *Farmer First*. Intermediate Technology Publications, London.

Dillon, J.L. and Hardaker, J.B. (1980) *Farm Management Research for Small Farmer Development*. FAO Agricultural Services Bulletin 41, FAO, Rome.

Jusoh, M. and Norton, G.A. (1987) Cabbage aphid control on commercial farms in the Thames Valley, UK. *Crop Protection 6*, 379–387.

Mascarenhas, J., Shah, P., Joseph, S., Jayakaran, R., Devavaram, J., Ramachandran, V., Fernandez, A., Chambers, R. and Pretty, J. (1991) *Participatory Rural Appraisal*. RRA Notes No. 13, International Institute for Environment and Development, London.

Mumford, J.D. and van Hamburg, H. (1985) *A Descriptive Analysis of Cotton Pest Management in South Africa*. Plant Protection Research Inst., Pretoria, 95 pp.

Siegel, S. (1956) *Nonparametric Statistics*. McGraw-Hill, New York.

Tull, D.S. and Hawkins, D.I. (1976) *Marketing Research*. Collier Macmillan International, New York.

USDA (1981) *The Delphi: Insecticide use and Lint yields, Beltwide Boll Weevil/Cotton Insect Management Programs. Overall Evaluation Appendix D*. ERS Staff Report No. AGESS 810507.

6 Introduction to Pest Models

G.A. NORTON[1], J. HOLT[2], AND J.D. MUMFORD[3]

[1]Cooperative Research Centre for Tropical Pest Management, University of Queensland, Brisbane, QLD 4072, Australia: [2]Natural Resources Institute, Central Avenue, Chatham Maritime, Chatham ME4 4TB, UK: [3]Department of Biology, Imperial College at Silwood Park, Ascot SL5 7PY, UK

INTRODUCTION

The term 'pest model' can strictly be applied to any representation of one or several processes associated with pest development and control, including conceptual models and the various graphic techniques illustrated in earlier chapters as descriptive tools. However, for our purposes here, a more restricted definition is used. We consider a pest model to be a mathematical, or at least computer-based, representation of a pest population, its development and mortality processes. A model may also include a pest's relationship with the crop or livestock host, and/or the processes involved in its control.

To illustrate the subject of modelling, this chapter sets out to answer four questions:

- Why model at all?
- What models can be used?
- How should models be designed?
- How are models tested?

WHY MODEL AT ALL (OR – HOW CAN MODELS BE USEFUL)?

The first point that needs to be made is that, for many problems in pest management, it is not necessary or sufficient to construct a pest model in order to improve pest management. In practice, improvements in pest management are often, if not mostly, achieved through the development of new technologies and management techniques, inspired empirical research, the reduction of major constraints, and by training farmers. However, as with the descriptive techniques described above, our argument is that models can **assist** the process of improving pest management. They can do this in a number of ways, some of the major reasons for modelling being summarized in the statements below:

1. Models represent explicit hypotheses of how key components and processes affect pest population development, crop damage, or the effectiveness of control. They are particularly useful where these hypotheses involve the interaction of a number of variables, which is usually the situation encountered in real-world systems.

2. Where independent data are available, a hypothesis — as a model — can be tested, leading to modification of the model and the identification of further key information needs.

3. Where data are not available, a model can possibly be even more useful. Consider the following four situations, in which models — as working hypotheses — are the only way in which these questions can be answered in a rigorous way:

A new pest has developed — there is little information available on the factors affecting the level of attack. What is the most critical information that should be collected?

There is a history of pesticide resistance having developed in a particular pest. How can current pesticides, to which resistance has not been detected, best be used to delay the onset of resistance?

What is the risk posed in a particular country by a quarantine pest that has not yet been introduced?

Biological control agents are to be sought for control of an introduced pest. How can we predict which is likely to be the most effective agent?

4. Models as working hypotheses can also provide:

Understanding of the underlying processes of pest population dynamics, crop–pest interactions, and effective strategies of control

A test-bed for experimentation, before carrying out expensive field trials.

Rigorous investigation of the performance of control strategies in specific situations, leading to the design of recommendations.

Predictions of pest attack and extrapolation from one situation to another.

Operational tools for real-time guidance to pest control decision making.

WHAT MODELS CAN BE USED?

Even a brief review of the use of models in pest management (for example, Norton and Holling, 1976; Conway, 1984; Getz and Gutierrez, 1982; Campbell and Madden, 1990) will illustrate that a range of modelling techniques can be, and have been, used. This is demonstrated in Table 6.1, which, at the same time, also provides a more immediate illustration of how models can be useful in pest management, both for research and operational purposes.

While the list of models set out in Table 6.1 is by no means comprehensive, it does give a good indication of the range of modelling possibilities available. They fall into three main categories:

Statistical models — such as regression models, which are based purely on empirical observations.

Mechanistic models — such as analytical and simulation models, which are constructed as representations of the basis of the underlying biological and ecological processes.

Optimization models — developed on the basis of procedures for finding optimal solutions.

Let us look briefly at each of these categories.

Table 6.1. Some modelling approaches in pest management.

Type of model	Examples
Regression models	Sugarbeet yellows prediction in the UK. Prediction of % yellows is based on the number of days with minimum temperature <0°C in Jan–March, and the mean temperature in April (Watson *et al.*, 1975)
Infection period models	Predict infection periods of disease on the basis of relative humidity and temperature requirements of the disease (Campbell and Madden, 1990)
Physiological (day-degree) models	'PETE' model in the USA. Pea moth and cutworm models in the UK – that start with trap catches and predict critical life stages for timing of control (Welch *et al.*, 1978)
Simulation models	Simulation of pest populations that allow control strategies to be 'tested', and of crop–pest relationships, to provide insights into damage relationships (see Chapter 8)
Analytical models	Models of predator–prey relationships that provide insights into the conditions in which biological control will be sustainable (see Chapter 7)
Optimization models	Use of Dynamic Programming to determine optimal control strategies for alfalfa weevil in the USA (Shoemaker, 1980)

Statistical Models

Statistical models, including various forms of regression analysis, offer a means of determining the value of a response variable using a small number of explanatory variables. They make no attempt to describe processes, although the terms included in the regression equation may say something about the nature of the relationship.

An example of a simple regression model is provided by a model (Fig. 6.1) that predicts the population density of brown planthopper (BPH) in the Yangtze Delta of China (Cheng and Holt, 1990). The magnitude of the population peak in the second generation (Y) is predicted by the magnitude of the population peak in the first generation (X) using the equation:

$$Y = 45.6738 \ X^{0.8895}$$

Although the relationship is empirical, based on historical data, the assumption of density-dependent population growth is implicit in the model. At lower densities, therefore, population growth is greater than would be the case at higher densities.

Another form of statistical model presents probabilities of outcomes, rather than the specific outputs given by regression models. Knight *et al.* (1992) illustrate this with an aphid model based on a database of outbreaks and aphid trap records (Fig. 6.2). For a given set of conditions a probability of attack level is presented, allowing the decision maker to see the risk of not controlling aphids.

One of the most well-known models used in practical crop protection is EPIPRE – Epidemic Prediction and Prevention – for wheat diseases in Europe (Zadoks, 1981). For

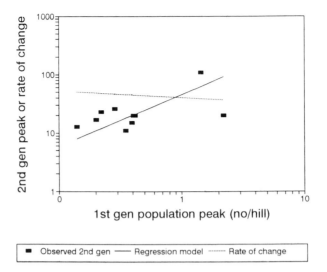

Fig. 6.1. Examples of a statistical model for brown planthopper in China (Cheng and Holt, 1990): a regression model to predict the second generation population peak from the first generation. The rate of change of the population between generations declines as density increases.

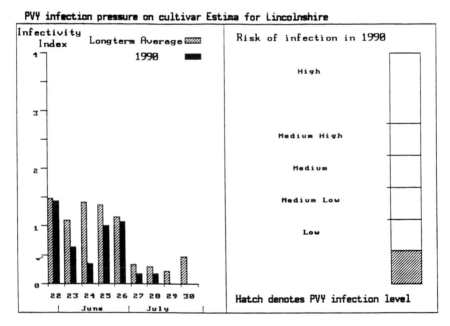

Fig. 6.2. Example output from FLYPAST PVY expert system showing aphid infection on seed potatoes. The left side of the figure shows the long-term average of the aphid infectivity index and that for the selected year for the critical months of June and July. The right side shows the level of risk to the crop that has been grown. In this example the risk is low and the crop would probably be suitable to be saved as seed (Knight *et al.*, 1992).

research purposes, various simulation models of the disease had been developed. However, for the practical purpose of advising farmers whether they should spray a particular crop, not spray it, or monitor again at a later stage, a linear regression model, that assessed likely disease development, was found to be more suitable.

Mechanistic Models

These models are constructed to represent deeper knowledge concerning the biological and ecological mechanisms or processes that underlie pest population dynamics, damage, and/ or control. Because they attempt to include biological realism, mechanistic models are likely to be most suitable for expressing hypotheses about the reasons for pest outbreaks, for example, or why certain biological control agents are likely to be more successful.

Mechanistic models can describe biological processes, such as the dynamics of populations in either continuous or discrete time. While models in continuous time should be a more realistic description of population processes when generations overlap, inclusion of point events (for example, a pesticide application) is more difficult. The alternative, models in discrete time, proceed in steps of time (days, hours, etc.) and fecundity, mortality, etc. are updated at each iteration.

The following chapters provide further discussion of the mechanistic type of modelling approach:

Chapter 7 – Analytical models
Chapter 8 – Simulation models
Chapter 9 – Rule-based models
Chapter 10 – Modelling with spreadsheets

Optimization Models

This type of modelling approach, as the name suggests, is primarily concerned with prescriptive decision problems in pest management. These models are constructed in such a way that the optimal – which often means the most profitable – strategy of pest management can be determined, either by finding a mathematical solution or by some numerical optimization procedures.

Dynamic Programming is a technique that is used to determine the best strategy to adopt in multi-stage decision-making problems, a category into which pest management usually falls. This technique has the unusual feature of starting at the latest point in time, say at harvest, and working backwards to the earliest decision point (Fig. 6.3). The advantage of this is that at each stage, once the optimal action plan for different levels of pest attack at that stage through to the end of the season has been determined, all other paths from that point can be ignored.

The analysis begins by determining the best strategy (that minimizes the combined costs of pest damage and cost of control) for the period t_3 to t_4. For the case shown in Fig 6.3; a_1 (1 spray) is the best treatment when the level of attack at t_3 is N_1. Similar calculations are carried out for each level of N at t_3, to determine the best strategy for each level of attack. All other strategies are inferior and are eliminated.

The analysis then moves back a step, to t_2 and determines the best strategy during

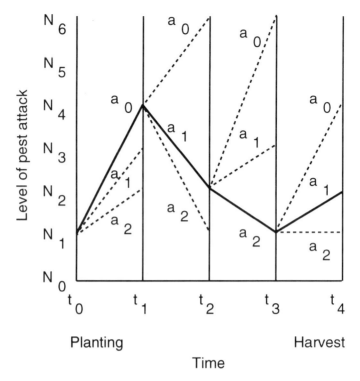

Fig. 6.3. A diagram illustrating the Dynamic Programming concept (a_0 = no control; a_1 = single spray; a_2 = two sprays). For further explanation see text.

the period t_2 to t_4. Starting at N_2 and t_2, the best strategy is a_2 and a_1 for the two periods. All other combinations at N_2, t_2 are inferior and are eliminated. Thus the analysis proceeds until optimal strategies are determined for each starting condition (N_0 to N_6 at t_0).

This technique has been applied to a number of pest problems, including alfalfa weevil (Shoemaker, 1984), spruce budworm (Holling *et al.*, 1976) and sugar cane froghopper (Conway *et al.*, 1975). The major problem with the technique is that it runs into difficulties if the system being analysed has more than two or three variables. This restriction often means that biological realism has to be sacrificed in order to undertake Dynamic Programming.

Other Models

As with any classification system, some models are unlikely to fit neatly into one or other of the three classes described above. To illustrate the point, CLIMEX — a climatic matching program — allows the user to combine summary data on climate (temperature, rainfall and relative humidity) for specific locations with information on the climatic tolerance of particular species, to determine the suitability of those locations for the particular species (Sutherst and Maywald, 1985). The various functions available in CLIMEX allow it to be used for a variety of purposes, including:

- Assessing the risk associated with pest introductions to new areas (Sutherst *et al.*, 1989).

- Deciding on appropriate areas to search for potential biocontrol agents during the exploratory phase of classical biological control (Worner *et al.*, 1989).
- Appraising the likely changes in pest status associated with climatic change scenarios (Sutherst, 1991).

A particular feature of CLIMEX is that it can be used in an empirical way (fitting parameters iteratively by matching simulated pest distributions with known distributions), or in a more analytical way (using information on the effect of climatic parameters on survival and development processes to assess pest suitability). This makes it difficult to accommodate CLIMEX in the simple classification above.

How Should Models Be Designed?

There are two important steps that should be followed in planning and designing models, whatever the context or the reason for modelling: problem specification and model specification.

Problem Specification

Since all models are most useful when directed at a clearly identified and well-defined problem, a crucial first step is to determine the question(s) that the model is to address. For example, is the model to be used for training, to act as a means of pulling together and interpreting existing research results, or to be used by practising pest managers as a means of decision support? If the latter, what is the purpose of the model, and is it concerned with:

- Predicting the timing of events, such as when disease infection is likely to occur, or when an insect pest is likely to reach a vulnerable stage for control.
- Predicting the scale of events, such as the size of disease infection at a certain growth stage, or the likely loss caused by an insect outbreak.
- Estimating the frequency or probability of events.
- Assessing and comparing the performance of different control strategies – the methods, timing, and combinations – in controlling pest populations.

In determining precisely what it is you wish to do with the model, the primary decision tool techniques, described in earlier chapters, have a crucial role to play in specifying the decision problem, and the questions that the model is required to answer (Fig. 6.4). They also have a key role to play in the second step, model specification.

Model Specification

This concerns the choice of modelling technique, such as an analytical or simulation model, and questions concerning the scope and detail to be included in the model. The problem facing the modeller is where to draw the boundary of the system – to determine the scope of the model – and in what detail to simulate the components to be included in the model. This problem is represented diagrammatically in Fig. 6.4. Three major compo-

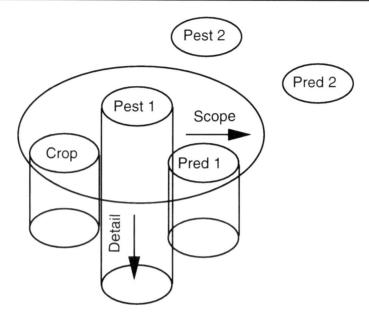

Fig. 6.4. Model specification should establish both the scope and the detail of the model.

nents are to be included in this particular model – crop, pest and predator – but at very different levels of detail. The detail on the crop component is very shallow. Perhaps a simple damage relationship is used to represent the impact of the pest on the crop, and a mortality coefficient is changed at a particular growth stage of the crop to represent the effect of the crop (in terms of food quality) on the pest. Detail on the pest is much deeper, reflecting the inclusion of (say) immigration, fecundity, and mortality processes in the model. Note that pest 2 and predator 2 are not included in the model.

In other cases, if a model is intended to examine a small set of discrete questions, or is to be used to investigate management options which have a relatively large impact on a pest population, then it might be argued that a model which describes pest dynamics in a relatively simple manner is all that is required, provided that the key processes in the dynamics of the pest are captured by the model.

In addition to the level of detail required to answer the specific question to be addressed by the model, another important factor that needs to be considered in model specification is the precision of estimates of values of the parameters and variables to be included in the model. There is little point in building a complex model that requires a level of precision in the estimation of parameters and variables that cannot be obtained in practice.

The following points must be kept in mind when developing a model specification:

- As scope and detail increase, the number of interactions to be described increases geometrically. This invariably means an increasing number of assumptions in the absence of experimental evidence. Parameterization and associated validation become very difficult – when the model fails to fit field results, any one of a large set of parameters might be the cause.
- Models of greater scope and detail may be more mechanistic – describing processes at a relatively deep level, so that interactions involving other elements of the system

are readily included. Models with few parameters may be more empirical, describing the data or the end-result rather than the process. More detail does not necessarily mean more realistic however, it may simply extend the boundary of the model, without improving its accuracy or explanatory capability. Despite the use of a large number of parameters, the processes captured by the model may still be simple.

- The appropriateness of the level of scope and detail in a model is difficult to judge because a single modelling approach is generally applied in any one situation. It is relatively easy to identify redundant detail in a model which is unnecessarily complex, less easy to identify wrong assumptions in a model which does not capture the key processes.

The implications of choices associated with system and model definitions are considered in more detail in later chapters. Typically, a number of considerations affect these choices:

Objectives of project – The range of questions which we wish the model to address and resources available for the modelling exercise should be central considerations.

Available data – In practice, models are often made as complicated and detailed as available data allow, rather than being governed by the objectives of the model.

Modelling philosophy – Different research groups have different modelling approaches, ranging from simple models with 'transparent' dynamics to very detailed, possibly more realistic models, the subtleties of which are difficult to validate.

Note that constructing a well-designed model is a relatively straightforward technical task and is not discussed in detail here. However, some examples are presented, particularly in Chapter 10 on spreadsheet models.

How Are Models Tested?

There are three main procedures used to test a model:

Verification – Do the equations and relations in the model correctly represent the system processes as intended? For instance, are there mistakes in the programming that mean the model does not actually do what it is intended to do? Some errors are obvious, but with others the model may superficially appear to function correctly. These errors are harder to detect. It is necessary to check each part of the program carefully for such errors, and to run the whole program on sample values for which the output is known.

Validation – Once a model has been verified (and so works as intended) it is necessary to compare the output from the model (for instance, pest population density) with independent field measurements of this variable for a range of places and times. Does the model provide a reasonable representation of the real system? Parameter values can be changed, either systematically or by trial and error, until a satisfactory representation of reality is achieved. This decision must be based on a predetermined level of accuracy that is related to the sort of decisions for which the model is intended. Coarse decisions (for instance, overall level of pest outbreak in a region) may not need a model that is as close to reality as fine decisions (for instance, spray timing in a specific field).

Sensitivity analysis – How does model output vary when model parameters are altered? It is important to know which elements of the system have the greatest impact on the output (in part because errors in these values would be most likely to cause problems with validity). A sensitive parameter is one which has a large impact on model output for a relatively small change in its own value. Interpretation of the relative impact of parameter changes in the real system must take into account the fact that some parameters are inherently more variable than others. A 'sensitive' parameter may actually show very little variation in nature.

CONCLUSION

In this introduction to models we have tried to show that models are useful for both understanding pest systems and for helping to guide action to solve pest problems. There is a wide choice of types of model and it is important to consider carefully the purpose of a model before developing one. Design is a critical, and often underestimated, stage in developing models. Both the problem and the model must have rigorous specifications to ensure that everyone concerned with a model's development knows what their targets are. Testing should be against predetermined levels of accuracy that are set by the decisions that will be made as a result of the model output.

REFERENCES

Campbell, C.L. and Madden, L.V. (1990) *Introduction to Plant Disease Epidemiology*. Wiley, New York.

Cheng, J. A. and Holt, J. (1990) A systems analysis approach to brown planthopper control on rice in Zhejiang Province, China. I. Simulation of outbreaks. *Journal of Applied Ecology* 27, 85–99.

Conway, G.R. (1984) *Pest and Pathogen Control: Strategic, Tactical, and Policy Models*. Wiley, New York.

Conway, G.R., Norton, G.A., Small, N.J. and King, A.B.S. (1975) A systems approach to the control of the sugar cane froghopper. In: Dalton, G.E. (ed.) *Study of Agricultural Systems*. Applied Science Publishers, London, pp. 193–229.

Getz, W.M. and Gutierrez, A.P. (1982) A perspective on systems analysis in crop production and insect pest management. *Annual Review of Entomology* 27, 447–466.

Holling, C.S., Jones, D.D. and Clark, W.C. (1976) Ecological policy design: A case study of forest and pest management. In: Norton, G.A. and Holling, C.S. (eds) *Pest Management – Proceedings of an International Conference*. Pergamon Press, Oxford, pp. 13–90.

Knight, J.D., Tatchell, G.M., Norton, G.A. and Harrington, R. (1992) FLYPAST. An Information Management System for the Rothamsted Aphid Database to Aid Pest Control Research and Advice. *Crop Protection* 11, 419–426.

Norton, G.A. and Holling, C.S. (1976) *Pest Management – Proceedings of an International Conference*, October 1976. Pergamon Press, Oxford.

Shoemaker, C.A. (1980) The role of systems analysis in integrated pest management. In: Huffaker, C.B. (ed.) *The New Technology of Pest Control*. Wiley, New York, pp. 25–49.

Sutherst, R.W. (1991) Pest risk analysis and the greenhouse effect. *Review of Agricultural Entomology* 79, 1177–1187.

Sutherst, R.W. and Maywald, G.F. (1985) A computerised system for matching climates in ecology. *Agriculture, Ecosystems and Environment* 13, 281–299.

Sutherst, R.W., Spradbery, J.P. and Maywald, G.F. (1989) The potential geographical distribution of the Old World screw-worm fly, *Chrysomia bezziana*. *Medical and Veterinary Entomology* 3, 273– 280.

Watson, M.A., Heathcote, G.D., Lauckner, F.G. and Sowray, P.A. (1975) The use of weather data and counts of aphids in the field to predict the incidence of yellowing viruses of sugar beet in England in relation to the use of insecticides. *Annals of Applied Biology* 81, 181–198.

Welch, S.M., Croft, B.A., Brunner, J.F. and Michels, M.F. (1978) PETE: An extension phenology modeling system for management of multi-species pest complex. *Environmental Entomology* 10, 425–432.

Worner, S.P., Goldson, S.B. and Frampton, E.R. (1989) Comparative ecoclimatic assessments of *Anaphes diana* (Hymenoptera: Mymaridae) and its intended host, *Sitona discoideus* (Coleoptera: Curculionidae), in New Zealand. *Journal of Economic Entomology* 82, 1085–1090.

Zadoks, J.C. (1981) EPIPRE: A disease and pest management system for winter wheat developed in the Netherlands. *EPPO Bulletin* 11, 365–396.

7 Analytical Models

T.H. JONES

Department of Biology, Imperial College at Silwood Park, Ascot SL5 7PY, UK

INTRODUCTION

Analytical models are those for which explicit formulae are derived for predicted values or distributions; they include regression and multivariate models, experimental designs, and the standard, and theoretical statistical distributions. Such models provide general information about an ecological system by describing its characteristics in a small number of relationships and parameters.

Most approaches to the use of analytical models have been based either on differential or difference equations. Differential models are appropriate for organisms with overlapping generations and more or less continuous births and deaths. This makes them suitable for conditions most frequently met in a tropical rain forest or a heated greenhouse. Organisms whose populations exhibit discrete generations and which show a succession of different developmental stages are best suited to difference equations. In these populations, events progress in a stepwise fashion, each step usually representing a generation.

Analytical models have been used as tools for developing pest management principles and ecological theory. This chapter explores their role, concentrating on models of predator–prey (parasitoid–host) relationships, that provide insights into the conditions in which biological control is sustainable. First, the use of simple analytical models in describing population changes is investigated. Models that describe host–parasitoid interactions are then considered and the role of difference models in practical biological control programmes discussed. Finally, more general examples are given of the role of analytical models in pest management.

SIMPLE POPULATION MODELS

Exponential Growth Model

Assume that a population has the potential to increase exponentially (Fig. 7.1a), as

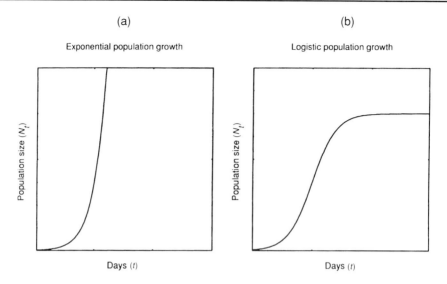

(a)   **(b)**

Exponential population growth   Logistic population growth

Fig. 7.1. Population growth curves. (a) Geometric growth in an unlimited environment and (b) logistic growth in a limited environment.

expressed by the equation

$$dN/dt = rN \tag{1}$$

where the rate of change in the numbers of population N with time, dN/dt, is the product of the size of the population (N) and its intrinsic growth rate (r). To find the population size at any time t, integration gives

$$N_t = N_o \exp^{rt} \tag{2}$$

where N_o is the population size at time t_o.

Logistic Growth Model

No population could sustain such an increase for long. Without other constraints, competition for resources would become increasingly severe and the net rate of increase dN/dt would be reduced, due to increased mortality, reduced fecundity, or both. This may be expressed mathematically as

$$dN/dt = rN(K - N)/K \tag{3}$$

where K is the carrying capacity – the maximum population size that the environment's resources can sustain. Thus, at low population numbers $(K - N)/K$ approximates 1 and growth approximates the exponential increase described above. However, as the population increases, there is more and more feedback from the term $(K - N)/K$. The population growth is now sigmoid (Fig. 7.1b) and the net rate of increase declines until, when $N_t = K$, there is no further change in population size $(dN/dt = 0)$ and the population is at its equilibrium level ($N^* = K$). The model is a stable one since the population will always tend to return to its equilibrium.

This logistic model (Verhulst, 1838) is a very simplified description of population change and several of the assumptions upon which it is based are not biologically realistic. It assumes, for instance, that there is an immediate effect of population size N on subsequent population change. Clearly, some time delays between cause and effect are inevitable: a population only reproduces at discrete intervals, in which case the effects of competition on natality are manifest only after the next breeding season. Time delays can also arise where a population exploits a resource which is then depleted for subsequent populations.

Such delays can have marked effects on the stability of populations – defined as the tendency to return to original (equilibrium) levels following a disturbance – and therefore should be included where appropriate. This can be done as follows:

$$dN/dt = rN_t(K - N_{t-T})/K \qquad (4)$$

where T is the time delay. In this case, the feedback between competition and the rate of population change depends upon population size at time $(t - T)$ rather than at t. Nevertheless, despite its simplicity, the logistic model can show, in a general way, the kinds of effects that density-dependent relationships can have on populations.

PREDATOR–PREY (PARASITOID–HOST) INTERACTIONS

Two main series of models have been developed to explain the dynamics of predator–prey (or parasitoid–host) interactions. One series uses differential equations to model predator–prey interactions. This approach was first developed by Lotka (1925) and Volterra (1926). The other series employs difference equations within discrete generations. These models are based on the work of Nicholson (1933) and Nicholson and Bailey (1935).

Lotka–Volterra Model of Predator–Prey Interaction

The Lotka–Volterra model (Fig. 7.2) describes the interaction between a prey species, with a density N, and its predator, with a density P, by the differential equations

$$dN/dt = rN[(K - N)/K] - c_1NP \qquad (5)$$

$$dP/dt = -dP + c_2NP \qquad (6)$$

where r is the prey's intrinsic rate of increase and d is the death rate of predators in the absence of prey. The constant c_1 may be interpreted as a coefficient of attack, and c_2 as a conversion factor of prey into more predator individuals.

In the absence of predation, the prey species follows the logistic equation, with an intrinsic rate r and a carrying capacity K. When predation does occur, the rate at which prey are eaten is proportional to the product of predator and prey densities (Equation 5).

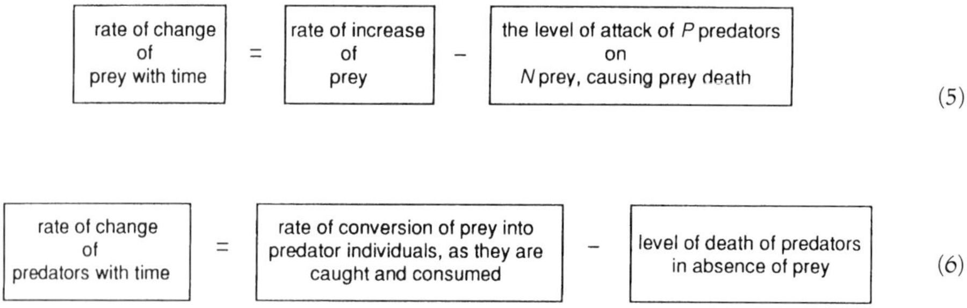

| rate of change of prey with time | = | rate of increase of prey | − | the level of attack of P predators on N prey, causing prey death | (5) |

| rate of change of predators with time | = | rate of conversion of prey into predator individuals, as they are caught and consumed | − | level of death of predators in absence of prey | (6) |

Fig. 7.2. Diagrammatic representation of the Lotka–Volterra model – normally expressed as a pair of differential equations.

Nicholson–Bailey Model of Parasitoid–Host Interaction

Most difference equation models for parasitoid action on host populations are elaborations of a basic form:

$$N_{t+1} = \lambda N_t f(N_t, P_t) \tag{7}$$

$$P_{t+1} = cN_t[1 - f(N_t, P_t)] \tag{8}$$

Equation 7 shows how host numbers change from one generation (N_t) to the next (N_{t+1}) as a function (f) of their own and parasitoid (P_t) number and behaviour, driven by λ, the host rate of increase per generation. In equation 8, c defines the mean number of parasitoid progeny produced per host attacked and $f(N_t, P_t)$ is the mortality function (a representation of the proportion of hosts surviving parasitism).

Assuming that parasitoids search for hosts at random, the Poisson distribution may be used to model the distribution of attacks between the hosts. Thus the proportion of hosts attacked at least once may be defined by $1 - \exp^{-aP}$ where a is the characteristic searching efficiency of a given parasitoid species – the Nicholsonian *area of discovery*. A population model for a host species (N) reproducing at rate λ per generation is obtained by substituting $\exp^{-aP}$, the proportion surviving, for $f(N_t, P_t)$, in Equation 7. This gives

$$N_{t+1} = \lambda N_1 \exp^{-aP_t} \tag{9}$$

where N_{t+1} is the number of hosts in the next generation. For the parasitoid

$$P_{t+1} = cN_t(1 - \exp^{-aP_t}) \tag{10}$$

The model leads to an unstable situation (Fig. 7.3). Nicholson (1933) suggested that although local extinctions are likely to occur, immigration and emigration would ensure the persistence of parasitoids and hosts over a more general area.

Stability in the Nicholsonian model can be brought about if the host population exhibits some form of density-dependent regulation through intraspecific competition. However, while such density-dependent regulation must always apply to populations faced with limited resources, host–parasitoid equilibria will occur at levels where this effect is not at all marked. It is therefore essential to look to the parasitoids themselves as a means of stabilizing interactions. To encapsulate parasitoid searching behaviour

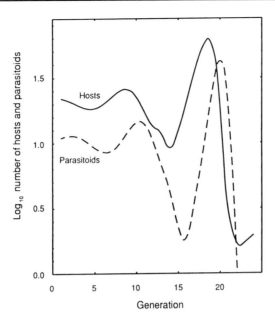

Fig. 7.3. Simulation produced by a population model based on Nicholson's (1933) theory; $a = 0.067$; $\lambda = 2$. (After Burnett, 1958.)

within a single constant is a gross simplification. The area of discovery $- a -$ is known to be an infinitely variable attribute and the way in which it varies may have a profound effect on the stability of the system.

The Nicholson–Bailey model assumes, implicitly, that the parasitoid exhibits a linear functional response. That is, as the density of hosts increases, the number of hosts attacked in a fixed period of time increases in proportion. Such a response (Type I) is likely to be typical of aquatic filter-feeding invertebrates that consume plankton in direct proportion to its surrounding abundance. Such feeding tends to cease abruptly when the animal is satiated, giving rise to a discontinuous plateau (Fig. 7.4a).

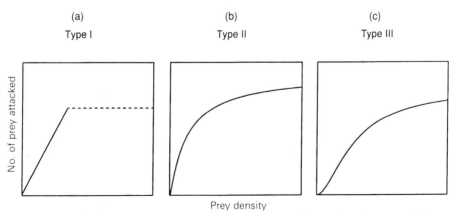

Fig. 7.4. Three types of functional response showing the changes in the number of prey attacked per unit time by a single predator as the initial prey density is varied. (After Holling, 1959.)

Type II responses (Fig. 7.4b) were thought, until recently, to be characteristic of invertebrate parasitoids and predators. The acts of quelling and egg-laying (killing and eating of prey) and then perhaps cleaning and resting, are all time-consuming activities (collectively known as 'handling time') which reduce the time available for search. Type II responses incorporate this reduction in searching time as host density increases. However, evidence is accruing (van Lenteren and Bakker, 1976; Hassell *et al.*, 1977; Jones, 1986) that Type III responses (Fig. 7.4c) are also widespread among arthropod parasitoids and predators. The Type III response, unlike both Type I and II responses, has an accelerating phase that is density dependent. In continuous, Lotka–Volterra type models, Murdoch and Oaten (1975) and Murdoch (1977) demonstrated that such a sigmoidal response may, in certain situations, have a stabilizing influence. Although Hassell and Comins (1978) found no such effect on discrete time Nicholson–Bailey models, Nunney (1980) has subsequently shown that if the parasitoid is able to discriminate parasitized hosts, sigmoid functional responses do have the potential to stabilize these models.

Hassell and Varley (1969) incorporated an expression for searching efficiency, $a = QP^m$, in the Nicholson–Bailey model that described the negative relationship that is often observed between log searching efficiency and log parasitoid (predator) density (Fig. 7.5). The mutual interference constant, m, can confer stability on the system. Q, which is Nicholson's a for one parasitoid (i.e. no interference), affects the equilibrium level but not stability.

Even if parasitoid searching efficiency is modified according to parasitoid and host density, the assumption that randomly searching parasitoids are foraging for randomly arranged hosts makes the model little different from that of Nicholson's. Consideration needs to be given to the effect of host density and distribution on parasitoid searching efficiency. One consequence of the response of parasitoids to host heterogeneity is its potential for inducing population stability. Some of the earlier theoretical literature emphasized the regulatory potential of direct density-dependent patterns of parasitism, resulting from parasitoids aggregating in patches of high host density (e.g. Hassell and

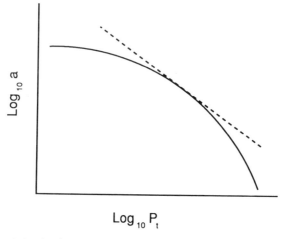

Fig. 7.5. Interference relationship between searching efficiency (expressed as the area of discovery, a) and parasitoid density per experiment, P_t, on logarithmic scales. (From Hassell and Varley, 1969.)

May, 1973, 1974; Murdoch and Oaten, 1975). More recently, it has been emphasized that, in addition to these direct density-dependent patterns, inverse density dependence (Hassell, 1984; Walde and Murdoch, 1988) and variation in parasitism that is independent of host density (Chesson and Murdoch, 1986; Hassell and May, 1988), can both be important contributors to population regulation if they involve sufficient heterogeneity in the risk of parasitism between individuals in the host population.

Most recently, Pacala *et al.* (1990) and Hassell *et al.* (1991) have shown how the effects of spatial heterogeneity of parasitism in a wide variety of discrete-generation host–parasitoid models can be assessed using a simple rule. Specifically, this rule states that the coefficient of variation squared ($CV^2 = [variance/mean^2]$) of the density of searching parasitoids in the vicinity of each host must typically exceed unity for the heterogeneity in parasitism to be a sufficient cause of stability in the interacting populations.

USE OF ANALYTICAL MODELS IN PEST CONTROL

Having briefly summarized the key features of predator–prey models, let us now consider how they might be of value in pest management.

A Theoretical Basis for Biological Control

First, we might ask the question – can theory provide a base for biocontrol attempts? This question is discussed further in Chapter 16, in which a holistic approach to the problem of biological control is suggested. The degree of depression of the host population below its carrying capacity was defined by Beddington *et al.* (1978) as:

$$q = N^*/K \tag{11}$$

where N^* is the equilibrium pest population level after the enemy's action and K is the maximum host population level reached in the absence of the parasitoid.

Despite the large number of biocontrol attempts recorded in the literature, Beddington *et al.* (1978) found it difficult to obtain values of N^* and K. This suggested that accurate before-and-after assessments of biocontrol attempts are rare. Secondly, although laboratory examples involved significant declines in host numbers, q in these cases was one or more orders of magnitude higher (i.e. less suppression) than in the field examples.

The identification of biologically useful and measurable life-history parameters in population models has proved difficult (Waage, 1990). One reason for this is that key parameters identified by analytical models, such as searching efficiency, handling time and aggregation are very difficult to measure in the field, while their measurement in the laboratory, although considerably easier, is often unrealistic. Nevertheless, it has been possible to explore the dynamic consequences of various parasitoid attributes within the framework of a simple model (Nicholson–Bailey) and to recommend criteria for the selection of biological control agents (Table 7.1; May and Hassell, 1988). Other things being equal, an ideal parasitoid for biological control in a perennial system emerges as one with a high search rate and a marked ability to aggregate in

Table 7.1. Parameters from predator/prey models and their contributions to stability, i.e. potential value in biological control. The information applies to monophagous predators/parasitoids in a more or less stable crop, e.g. a forest. (From Putman and Wratten, 1984.)

Parameter	Optimum for biological control	Effects and limitation of parameter
a',a Attack rate Q	High or	Reduces average population level; no effect on stability
$*a'$	a' changes with prey density	Sigmoid functional response – gives stability only in models with no time-delays (e.g. in differential equations) but not in models with time-delays (e.g. difference equations)
T_t Total searching and feeding time	High	Reduces average population level; no effect on stability
T_h Handling time as proportion of T_t	Low	Slight increase in stability; slight increase in average population level
$*m$ Mutual interference constant (Hassell and Varley, 1969)	0–1	Optimum value depends on prey reproductive rate (F or λ) – increase in stability but large values of q only. Field relevance?
$*\mu$ Predator aggregation index (Hassell and May, 1973)	High	0 = random search, ∞ = all predators in one patch. Increase in stability but influenced by α (proportion of prey in the one large patch) and λ (prey rate of increase)
$*k$ Predator aggregation index (k = exponent of negative binomial distribution (Hassell and May 1974))	All values of $k < 1$ (range is 0 to ∞)	Stability but needs some prey intraspecific density dependence if $k \ll 1$

* = parameters giving stability.

patches of high host density. The high search rate promotes low equilibrium host populations and the aggregation is necessary for this equilibrium to be stable. This simple model may be enhanced by including further functions. Two specific examples are discussed below: the role of generalist and specialist natural enemies, and the effect of insecticide application on host–parasitoid interactions.

Generalist and specialist natural enemies in insect predator–prey interactions

Hassell and May (1986) considered the dynamics of a predator–prey, or parasitoid–host, interaction where the predator or parasitoid is a generalist whose population is buffered against changes in the particular prey being considered. They then broadened the interaction to include, in addition, a specialist natural enemy. They asked:

- Under what conditions can a specialist invade and persist in an existing generalist–prey interaction?
- How does the addition of the specialist natural enemy alter the population dynamics?
- How does the relative timing of specialist and generalist in the prey's life cycle affect the dynamics of the interaction?

The following conclusions emerged:

- A specialist can invade and coexist more easily if acting before the generalist in the prey's life cycle.
- In some cases the establishment of a specialist leads to higher prey populations than existed previously with only the generalist acting.
- A three-species stable system can readily exist where the prey–generalist interaction alone would be unstable or have no equilibrium at all.
- A variety of alternative stable states are possible, either alternating between two-species or three-species states, or between different three-species states.

Modern methods of analysis are making it easier to parameterize relatively simple population models from detailed field studies on particular systems and this is opening the way to a much closer integration of empirical and theoretical ecology. Jones *et al.* (1993) examined the relative contributions to the dynamics of the cabbage root fly, *Delia radicum* (L.) (Diptera: Anthomyiidae), of the spatial and temporal patterns of mortality caused by generalist and specialist parasitoid species. The resulting population model was parameterized by finding the maximum of a likelihood function for the 9 years' data reported in the study. Exploration of the model illustrates how:

- The specialist parasitoid, *Trybliographa rapae* (Westw.) (Hymenoptera: Cynipoidea), can promote stability by virtue of its spatial response to host density, but only within a very narrow range of host rates of increase, λ.
- By contrast, *Aleochara bilineata* Gyll. (Coleoptera: Staphylinidae), the generalist natural enemy, acts as a simple, between-generation density-dependent factor, tending to regulate the host population. As λ increases, stable host populations give way to locally unstable ones, exhibiting limit cycles or higher order behaviour.
- Provided that the survivorship of *T. rapae* is sufficiently high, the interplay of the two natural enemies can lead to alternative stable states although the population levels involved lie largely outside the range of observed densities from the field. These patterns are similar to those found by Hassell and May (1986) in their theoretical study of the role of generalist and specialist natural enemies on insect population dynamics. However, in that case, the model was not tied to a particular interaction and the parameter combinations required for alternative three-species states, of the kind found in the cabbage root fly model, required rather extreme values of λ.

Pest–parasitoid–insecticide interactions

Waage *et al.* (1985) extended the use of analytical host–parasitoid models to include insecticide application, allowing the impact of insecticides to affect pest hosts and/or

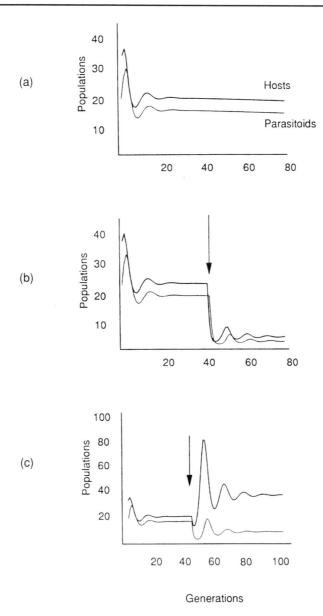

Fig. 7.6. Simulations of model predictions for the interaction of pests, parasitoids and insecticides (after Waage *et al.* 1985). (a) The underlying analytical parasitoid–host model, with hosts and parasitoids introduced in the first generation at population levels indicated; (b) the same, with pesticide application commencing at generation 40 (arrow) and killing hosts surviving parasitism but not affecting parasitoids; (c) the same, with pesticide application commencing at generation 40 (arrow) and killing adult parasitoids as well as parasitized and healthy hosts. (From Waage, 1989; courtesy of Intercept Ltd.)

their parasitoids in a variety of ways. These models were then analysed to show the extent of depression or resurgence resulting from insecticide use. Insecticides acting on the pest alone always contribute to host depression (Fig. 7.6b), to a degree dependent on whether they precede or follow parasitism and on the stage of the pest life cycle where equilibrium population size is measured.

When insecticide application also leads to the death of parasitoids, pest depression is reduced and pest resurgence may or may not appear (Fig. 7.6c), depending on the timing of application. Waage *et al.* (1985) showed that the relative toxicity of insecticides to pests, immature and adult parasitoids is crucial to the level of resurgence obtained. The aim of this modelling exercise was not to solve problems in integrated pest management – the approach adopted is too simplistic. Rather, as Waage (1989) states, the analytical modelling of pest–pesticide–natural enemy interactions helps identify a spectrum of possible dynamical outcomes ranging from resurgence, through self-cancelling effects of pesticides on pests and natural enemies, to additivity and even synergism.

Pesticide Resistance

Another example of the use of simple models to explore generic pest management problems can be found in the area of pesticide resistance. Comins (1977a) developed a deterministic Mendelian model, based on a number of simplifying assumptions. The pest population is assumed to be large and homogeneous with discrete generations, and to exhibit density-dependent survival of the form:

$$N_{t+1} = \lambda N_t^{1-b} \tag{12}$$

where N_t and N_{t+1} are numbers of pests in succeeding generations, λ is the population growth rate and b is the density-dependent coefficient (May *et al.*, 1974).

The resistance gene is assumed to operate equally in both sexes and develop in three phases (Fig. 7.7). Initially, when the pesticide is not being applied, the resistance gene is likely to be a liability and will exist at a low equilibrium frequency determined by the opposing forces of mutation and adverse selection. In the second phase, once pesticide selection has commenced, the resistance gene spreads rapidly through the population although still below the level at which control efficiency is noticeably reduced. Finally, control measures fail, resistance is detected and after a period of increasing dosages, the pesticide is abandoned.

A further assumption is made that the genes being selected exist initially at low frequencies. Thus, in the second phase, partially or fully dominant genes will be selected on the basis of the degree of resistance they confer in heterozygous individuals and this will be the most important factor determining the rate of development of resistance.

For perfect density dependence ($b = 1$) resistance time decreases with increasing kill, but where the relationship is undercompensating ($0 < b < 1$) intermediate kills result in the fastest resistance. In general, for reasonable levels of density dependence it seems that resistance can be suppressed by using a highly effective non-residual insecticide, strictly confined to the target population.

Further development of the model (Comins, 1977b) demonstrated the importance of pest migration in delaying resistance. If the treated pest population in a farmer's field interbreeds with a large untreated population in a wild habitat or in adjacent fields, then resistance in the treated population will develop more rapidly if the rate of migration from the untreated to the treated area falls below a certain critical threshold. Suscep-

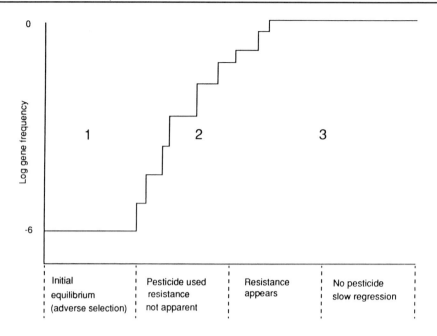

Fig. 7.7. Phases in the development of pesticide resistance. (After Comins, 1986.)

tible individuals migrating in provide 'susceptible' genes for the next generation and hence in effect 'dilute' the inheritance of resistance from the survivors in the treated area. Where the resistance gene is recessive the critical threshold is very sharply defined, the population suddenly jumping from a very low to a very high frequency of resistance. Even where the resistance genes are partially dominant, the rate of development of resistance will accelerate as migration decreases.

In general any strategy which increases the effective migration rate or the effective relative size of the untreated population is to be recommended. Other considerations aside, it is good policy to leave untreated subpopulations of pests in neighbouring fields or on wild host plants. It also follows that for pests with discrete generations pesticides should be withheld in the period between immigration and reproduction, so producing the maximum degree of reproductive competition.

Infectious Disease Agents

The role of analytical models in the understanding of infectious diseases has been fully reviewed and discussed by Anderson and May (1991). Population models of disease transmission have traditionally been couched in differential equation form. It is generally assumed that birth, death and infection processes are continuous where generations overlap completely.

Two distinct mathematical frameworks have been employed (Anderson and May, 1979). First, models of human infection are usually compartmental in structure, where the host population is assumed to be of constant size and interest is focused on the flow of hosts between compartments containing, for example, susceptibles, infecteds and immunes. The disease agents studied are usually microparasites – viruses, bacteria and protozoa – organisms characterized by small size, short generation times, and the ability to multiply directly and rapidly within the host. A second type of model has also been described in the ecological literature, where attention has recently been given to the population dynamics of host–parasite associations with particular emphasis on the way macroparasites – helminths and arthropods – depress the natural intrinsic rate of animal populations.

Consider, for example, the dynamics of a directly transmitted viral microparasite, such as measles, within a host population of constant size N. This population consists of susceptibles, infecteds and immunes (Fig. 7.8) numbering respectively, $X(t)$, $Y(t)$ and $Z(t)$ at time t. The per capita birth rate is a and is independent of whether the individual is susceptible, infected or immune, thus the net birth rate is $a(X + Y + Z)$. The natural mortality rate is b, and the rate at which infection is acquired is proportional to the number of encounters between susceptibles and infecteds, being βXY where β is some transmission coefficient. The mortality rate for infected individuals is taken to be $b + \alpha$, where α represents the mortality caused by the disease. There is also a recovery rate μ. Recovered individuals are initially immune but this immunity can be lost at a rate γ. These assumptions lead to the following differential equations:

$$dX/dt = a(X + Y + Z) - bX - \beta XY + \gamma Z \qquad (13)$$

$$dY/dt = \beta XY - (\alpha + b + \mu)Y \qquad (14)$$

$$dZ/dt = \mu Y - (b + \gamma)Z \qquad (15)$$

Using this as a general model, Anderson and May (1981) considered a series of variations

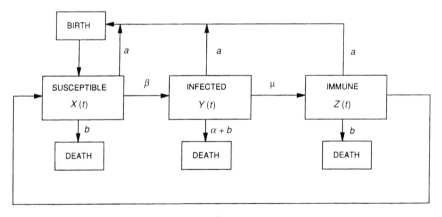

Fig. 7.8. Diagrammatic flow chart for a directly transmitted infection, described by a compartment model with susceptible (X), infected (Y), and immune (Z) hosts. The flow of individual hosts between compartments is controlled by a set of rate parameters: per capita birth rate, a; natural death rate of hosts, b; disease-induced mortality, α, acting on infected hosts; recovery rate, μ; transmission rate per encounter between susceptible and infected hosts, β; rate of loss of immunity, γ. Reprinted with permission from *Nature* (Anderson and May, 1979). Copyright (1979) Macmillan Magazines Limited.

Table 7.2. Summarized results of modifications made to the generalized host–disease model and consequences for interacting populations. (From Anderson and May, 1981.)

	Model modifications	Results
A	General model (no modification)	Stable equilibrium point OR exponentially increasing host with increasing proportion infected
B	Reduction in host reproduction	Same as A
C	Vertical transmission	Same as A OR endemic disease without regulation
D	Latent period	Same as A OR stable cycles
E	Virulence related to host density	Stable equilibrium point OR extinction of disease
F	Density-dependent mortality	Extinction of disease OR stable equilibrium below carrying capacity
G	Free-living infective stages	Stable equilibrium point OR stable cycles OR exponentially increasing host with disease OR extinction of disease
H	Model G plus seasonal growth	Similar to G OR seasonal OR non-seasonal cycles OR other extraordinarily complex behaviour

similar to those performed on the Nicholson–Bailey model, as described earlier. The results of this series of analytical models are summarized in Table 7.2.

Are these models realistic? All of the models lack some realism because Anderson and May (1979, 1981) decided not to include more than a few processes in any one model even though they knew about many processes that could be important. Therefore, how much value do these simple frameworks have? One advantage they do have is in being small, all equations can be studied simultaneously. On the other hand, their heuristic value seems to be constrained by two characteristics. First, the limited number of processes that can be included in any single model means that any one model and often any one report concerning this type of model has limited value because of their reduced realism. Often we need to read all of an analytical modeller's papers to gain a holistic view of all the conditions and processes that can affect a population or community. Second, because the coefficients in simple analytical models are highly aggregated, and do not reveal many underlying assumptions, the processes and interactions expressed in these models are often difficult to conceptualize and empirically estimate with statistical confidence. Anderson and May (1981), for example, admit that their single coefficient disease transmission rate is difficult to measure.

Although there now exists a well-developed theoretical framework for the study of the dynamics of infectious agent transmission and control, the value of this framework is not always well appreciated by decision makers. Fully aware of this, Anderson (1991) stresses that if such theory is to play a role in the solution of practical problems a much greater emphasis must be placed on data-orientated studies and the careful comparison of prediction with observation.

Using a similar approach to that described above, Hochberg (1989) has investigated the potential role of pathogens in biological control. By developing a model of invertebrate host–pathogen interactions incorporating the range of dynamics observed in natural systems Hochberg found that host populations may be regulated to low and relatively constant densities if 'sufficient numbers of pathogens are translocated from pathogen reserves to habitats where transmission can occur'. Thus he argues that an understanding of pathogen reservoirs may be of value in the design of biological control programmes and may greatly increase the effectiveness of pathogens as biological control agents.

Conclusions

A major dilemma in analytical modelling is deciding whether to create a simple model, and risk losing biological realism, or to create a complex model and lose generic value. Waage (1986) identified several problems with models that are simple, general, and mathematically elegant. He points out that there are more models than studies to test them, and that the models commonly make predictions that are too broad for confident testing because the similarity of the predictions prevents us from refuting any of them in an empirical study. These criticisms are valid to the majority of simple, analytical models. However, as May (1974) states, 'general models, even though they do not correspond in detail to any single real community, aim to provide a conceptual framework for the discussion of broad classes of phenomena. Such a framework can serve a useful purpose in indicating key areas or relevant questions for the field and laboratory ecologist, or simply in sharpening discussion of contentious issues.'

References

Anderson, R.M. (1991) Populations and infectious diseases: ecology or epidemiology? *Journal of Animal Ecology* 60, 1–50.

Anderson, R.M. and May, R.M. (1979) Population biology of infectious diseases. I. *Nature* 280, 361–367.

Anderson, R.M. and May, R.M. (1981) The population dynamics of microparasites and their invertebrate hosts. *Philosophical Transactions of the Royal Society, London, Series B* 291, 451–523.

Anderson, R.M. and May, R.M. (1991) *Infectious Diseases of Humans: Dynamics and Control*. Oxford Science Publications, Oxford.

Beddington, J.R., Free, C.A. and Lawton, J.H. (1978) Modelling biological control: on the characteristics of successful natural enemies. *Nature* 273, 513–519.

Burnett, T. (1958) A model of host–parasite interaction. *Proceedings Tenth International Congress of Entomology* 2, 679–686.

Chesson, P.L. and Murdoch, W.W. (1986) Aggregation of risk: relationships among host–parasitoid models. *American Naturalist* 127, 696–715.

Comins, H.N. (1977a) The management of pesticide resistance. *Journal of Theoretical Biology* 65, 399–420.

Comins, H.N. (1977b) The development of insecticide resistance in the presence of migration. *Journal of Theoretical Biology* 64, 177–197.

Comins, H.N. (1986) Tactics for resistance management using multiple pesticides. *Agriculture, Ecosystems and Environment* 16, 129–148.

Hassell, M.P. (1975) Density dependence in single-species populations. *Journal of Animal Ecology* 44, 283–295.

Hassell, M.P. (1984) Parasitism in patchy environments: inverse density dependence can be stabilising. *Institute of Mathematics and its Applications. Journal of Mathematics Applied in Medicine and Biology* 1, 123–133.

Hassell, M.P. and Comins, H.N. (1978) Sigmoid functional responses and population stability. *Theoretical Population Biology* 12, 62–67.

Hassell, M.P. and May, R.M. (1973) Stability in insect host–parasite models. *Journal of Animal Ecology* 42, 693–726.

Hassell, M.P. and May, R.M. (1974) Aggregation in predators and insect parasites and its effect on stability. *Journal of Animal Ecology* 43, 567–594.

Hassell, M.P. and May, R.M. (1986) Generalist and specialist natural enemies in insect predator–prey interactions. *Journal of Animal Ecology* 55, 923–940.

Hassell, M.P. and May, R.M. (1988) Spatial heterogeneity and the dynamics of parasitoid–host systems. *Annales Zoologici Fennici* 25, 55–61.

Hassell, M.P. and Varley, G.C. (1969) New inductive population model for insect parasites and its bearing on biological control. *Nature* 223 1133–1136.

Hassell, M.P., Lawton, J.H. and Beddington, J.R. (1977) Sigmoid functional responses by invertebrate predators and parasitoids. *Journal of Animal Ecology* 46, 249–262.

Hassell, M.P., May, R.M., Pacala, S.W. and Chesson, P.L. (1991) The persistence of host–parasitoid associations in patchy environments. I. A general criterion. *American Naturalist* 138, 568–583.

Hochberg, M.E. (1989) The potential role of pathogens in biological control. *Nature* 337, 262–265.

Holling, C.S. (1959) Some characteristics of simple types of predation and parasitism. *Canadian Entomologist* 91, 385–398.

Jones, T.H. (1986) Patterns of parasitism by *Trybliographa rapae* Westw., a cynipid parasitoid of the cabbage root fly. Unpublished PhD thesis, University of London.

Jones, T.H., Hassell, M.P. and Pacala, S.W. (1993) Spatial heterogeneity and the population dynamics of a host–parasitoid system. *Journal of Animal Ecology* 62, 251–262.

Lotka, A.J. (1925) *Elements of Physical Biology*. Williams and Wilkins, Baltimore. (Reissued as *Elements of Mathematical Biology* by Dover, 1956.)

May, R.M. (1974) *Stability and Complexity in Model Ecosystems. Monographs in Population Biology* 6. Second edition. Princeton University Press.

May, R.M. and Hassell, M.P. (1988) Population dynamics and biological control. *Philosophical Transactions of the Royal Society, London, Series B* 318, 129–169.

May, R.M., Conway, G.R., Hassell, M.P. and Southwood, T.R.E. (1974) Time delays, density-dependence and single-species oscillations. *Journal of Animal Ecology* 46, 747–770.

Murdoch, W.W. (1977) Stabilising effects of spatial heterogeneity in predator–prey systems. *Theoretical Population Biology* 11, 252–273.

Murdoch, W.W. and Oaten, A. (1975) Predation and population stability. *Advances in Ecological Research* 9, 2–131.

Nicholson, A.J. (1933) The balance of animal populations. *Journal of Animal Ecology* 2, 132–178.

Nicholson, A.J. and Bailey, V.A. (1935) The balance of animal populations. Part I. *Proceedings of the Zoological Society of London* 1935, 551–598.

Nunney, L. (1980) The influence of the Type 3 (sigmoid) functional response upon the stability of predator–prey difference models. *Theoretical Population Biology* 18, 257–278.

Pacala, S.W., Hassell, M.P. and May, R.M. (1990) Host–parasitoid associations in patchy environments. *Nature* 344, 150–153.

Putman, R.J. and Wratten, S.D. (1984) *Principles of Ecology*. Croom Helm, London, 388 pp.

van Lenteren, J.C. and Bakker, K. (1976) Functional responses in invertebrates. *Netherlands Journal of Zoology* 26, 567–572.

Varley, G.C., Gradwell, G.R. and Hassell, M.P. (1973) *Insect Population Ecology: An Analytical Approach*. Blackwell Scientific Publications, Oxford.

Verhulst, P.F. (1838) Notice sur le loi que la population suit dans son accroissement. *Correspondence Mathematique et Physique* 10, 113–121.

Volterra, V. (1926) Variazioni e fluttuazioni del numero d'individui in specie animali conviventi. *Memorie della Royale Accademia Nazionale dei Lincei* 2, 31–113. (Translation in Chapman, R.N. (1931) *Animal Ecology*. McGraw-Hill, New York, pp. 409–448.)

Waage, J.K. (1986) Family planning in parasitoids: adaptive patterns of progeny and sex allocation.

In: Waage, J.K. and Greathead, D. (eds) *Insect Parasitoids*. 13th Symposium of the Royal Entomological Society of London. Academic Press, London, pp. 63–96.

Waage, J.K. (1989) The population ecology of pest–pesticide–natural enemy interactions. In: Jepson, P.C. (ed.) *Pesticide and Non-target Invertebrates*. Intercept Ltd., Andover, UK, pp. 81–94.

Waage, J.K. (1990) Ecological theory and the selection of biological control agents. In: MacKauer, M., Ehler, L.E. and Roland, J. (eds) *Critical Issues in Biological Control*. Intercept Ltd, Andover, UK, pp. 135–158.

Waage, J.K., Hassell, M.P. and Godfray, H.C.J. (1985) The dynamics of pest–parasitoid–insecticide interactions. *Journal of Applied Ecology* 22, 825–838.

Walde, S.J. and Murdoch, W.W. (1988) Spatial density dependence in parasitoids. *Annual Review of Entomology* 33, 441–466.

8 Simulation Models

J. HOLT[1] AND G.A. NORTON[2]

[1]Natural Resources Institute, Central Avenue, Chatham Maritime, Chatham ME4 4TB, UK: [2]Cooperative Research Centre for Tropical Pest Management, University of Queensland, Brisbane, QLD 4072, Australia

INTRODUCTION

The purpose of this chapter on simulation modelling is not to teach the reader how to build simulation models but rather to provide some understanding of the nature and role of this modelling technique in pest management decision making. The majority of simulation modelling applications in pest management involve the simulation of the population dynamics of crops, pests, natural enemies of pests, or combinations of these. In this context, the chapter sets out to answer three questions:

- What is a simulation model?
- What kinds of problem might benefit from simulation analysis?
- What decisions need to be made about the type of simulation model to build?

Illustrations relating to these questions are provided by examples of different types of model.

WHAT IS A SIMULATION MODEL?

To illustrate the basic principles that lie behind simulation modelling, let us consider two particular pests, a disease, apple mildew, and an insect pest, the rice brown planthopper (BPH).

Apple Mildew

Mildew is a major pest of apples in the United Kingdom, Europe and North America. Its development on the crop over the season is dependent on the favourability of that season, particularly the number of rainy days, that affect leaf wetness. However, unlike many other plant pathogens, mildew does not have distinct infection periods; infection

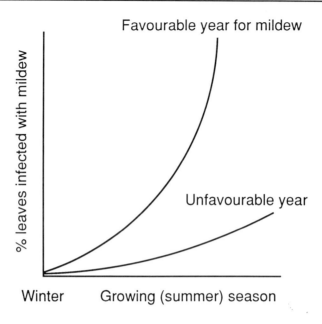

Fig. 8.1. A simple simulation model of mildew development on apples.

can be occurring all the time, at a faster or slower rate depending on the weather conditions.

During a workshop in 1978 (Barlow et al., 1979), a key question regarding mildew control was – what level of control will a novel winter fungicide treatment have to achieve for it to be a more cost-effective method compared with conventional calendar spraying in the summer?

As one means of exploring this question, a very simple simulation model of disease development was constructed. Inspection of field data showed that actual disease development could be reasonably represented by an exponential curve, as shown in Fig. 8.1.

In favourable years, the rate of increase of this curve is much higher than in unfavourable years. Expressed in mathematical terms, the development of mildew can be modelled in terms of the increment of disease development that occurs each day. Thus, the percentage of leaves infected with mildew (l) on day $t + 1$ equals the amount of disease on the previous day, multiplied by the rate of growth (r), which can be varied to represent a favourable or unfavourable season. That is:

$$l_{t+1} = l_t\, r$$

Thus, in the computer simulation, each day the amount of disease increases according to the equation above. To investigate management questions, like the one raised above, the effect of different fungicide treatments can also be simulated. For instance, when an eradicant summer spray treatment is given, say on day t, it causes a 60% reduction. Consequently, the level of infection on the following day ($t + 1$) will now be:

$$l_{t+1} = l_t(1 - 0.6)r$$

where $(1 - 0.6)$ = the proportion of infection remaining following the eradicant spray.

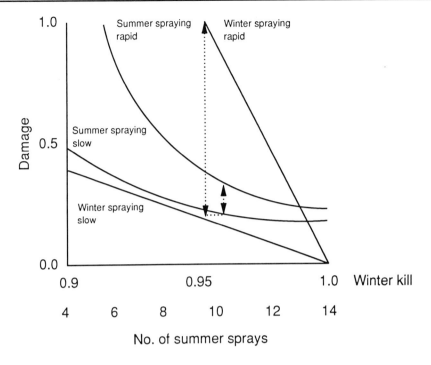

Fig. 8.2. A comparison of winter and summer spraying of mildew on apples given two rates of disease development. Damage is relative to the amount of overwintering mildew, and summer sprays are assumed to be eradicants applied at regular intervals over the time period considered, each causing a 60% mortality. 'Rapid' and 'slow' indicates rates of development of the infection dependent on weather; 'slow' denotes a ten-fold increase and 'rapid' a 100-fold increase over the time period. Vertical dotted lines show the increase in damage when the rate of disease development changes from slow to rapid, under winter and summer spraying programmes given initially similar damage. (After Barlow *et al.*, 1979.)

Using the simple simulation model in this way, the result shown in Fig. 8.2 was obtained. While a winter spray can be very effective in reducing total levels of infestation in the following season, a kill of virtually 100% needs to be obtained if it is to be an improvement on summer spraying, particularly when the season is favourable. By changing the components in the model, for instance, '*r*' and the percentage efficacy of the fungicide, the relative importance of these different components can be explored.

Brown Planthopper (BPH)

Now let us look at a more complex simulation problem, that of simulating brown planthopper (BPH), a major pest of rice in Asia. As we will see later in this chapter, a simulation model of BPH dynamics has been developed to improve our understanding of the factors affecting BPH outbreaks, and to assess the effect of different control measures.

The components and operation of the BPH simulation model are shown in Fig. 8.3. Each box represents the numbers of individuals in a particular age class, for a particular life stage, on a particular day. For instance, N_1 and N_2 represent the numbers of 0- to 1-day old and 1- to 2-day old nymphs. Each day, a number of processes are occurring:

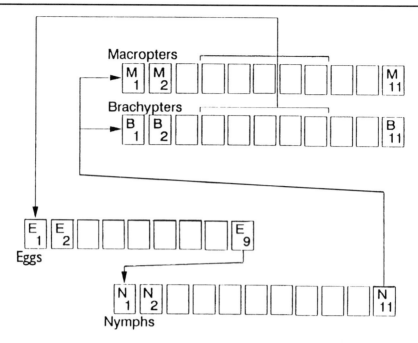

Fig. 8.3. BPH simulation model (for the Philippines). At each daily iteration, individuals are transferred from one daily age class to the next (further details in text).

1. A proportion of individuals are transferred from one box to the next, representing the ageing process.
2. The remaining proportion of individuals disappear from the system either in death, or as emigrants.
3. Eggs are laid by macropterous and brachypterous adults, between the ages of 3 to 4 and 7 to 8-days old, to produce the total number of 0- to 1-day old eggs.
4. Eggs that are 9–10 days old hatch into nymphs.
5. Nymphs that are 10–11 days old become macropterous or brachypterous adults.

While Fig. 8.3 is an accurate representation of how a simulation model works, to incorporate this in a computer program, these processes are expressed in the form of equations. For instance, the ageing process is very simply simulated: each day, individuals in one age class move to the next age class. Thus, for eggs, each day the number of 0- to 1-day old eggs (E_1) on the previous day becomes the number of 1- to 2-day old eggs (E_2) today, that is:

$$E_{2,t+1} = E_{1,t}$$

Similarly, for the number of 2- to 3-day old eggs:

$$E_{3,t+1} = E_{2,t}$$

However, this assumes that all eggs move to the next age class. As we noted in the list above (in points **1** and **2**), a proportion of eggs can die. In the BPH model, the proportional daily mortality of eggs is assumed to be a constant (e). Thus, we need to modify the equations above to include this mortality factor, as shown below:

$$E_{2,t+1} = E_{1,t}(1 - e)$$

and

$$E_{3,t+1} = E_{2,t}(1 - e)$$

In a similar way, the process described in **3** above can be represented as follows:

$$E_{1,t} = \sum_{i=4}^{8} [A_i(1 - a)f]$$

where A_i = number of adults of age i, i = 4 to 8 days old; a = the daily mortality of adults, assumed to be constant; f = the fecundity, or number of eggs produced per adult per day.

Where mortality of individuals is caused by factors such as predation and parasitism, this can be incorporated in the model in two ways. First, by representing mortality as a function of the density of the pest (prey), as described in more detail later in this chapter. Second, a more explicit approach can be adopted by simulating the predator or parasite population and directly simulating the prey caught or parasitized, as described in the previous chapter. This will be considered in more detail later in this chapter and in the subsequent chapter.

Another important consideration is whether the parameter values to be included in the model are input as constants (deterministic model) or as a probability distribution (stochastic model). Deterministic models always yield the same answer for a given set of inputs. In stochastic models, values of particular variables, such as development time, are drawn from a distribution so that each time the simulation model is run a different answer is obtained. Simulation results have a distribution which may reflect the variance found in nature but only if the variance of the individual model components is known.

Having determined how key biological processes can be represented in mathematical terms, there are various ways in which these models can then be computerized. The more traditional way is to use a procedural language, such as Fortran, Pascal, Basic, etc.. The advantage of this approach is that the program can be designed precisely to meet the users' needs. However, it can be very time consuming. Partly in response to this problem, special simulation languages, such as CSMP (de Wit and Goudriaan, 1978), have been developed, freeing the user from a good deal of the programming.

More recently, other reduced-programming alternatives have become available, through the use of spreadsheets and rule-based (expert system) shells. Using these programs, the modeller can be less concerned with the details of programming and can concentrate on the design of the model, and the constituent relationships and processes. The advantages of both these methods are described in Chapters 9 and 10. A more recent development (in the Cooperative Research Centre for Tropical Pest Management in Australia) is the production of a generic insect model, written in C++ and operating in

Windows. It enables non-programmers to build simulation models for specific pests. This is achieved by presenting the user with a series of questions set up as menus. As the user answers these questions, the structure and parameters of the model are incorporated in software modules.

WHAT KINDS OF PROBLEM MIGHT BENEFIT FROM SIMULATION ANALYSIS?

Most of the ways in which simulation models can provide a valuable contribution to pest management fall into one or more of the following three categories:

Understanding pest population dynamics and damage – Models can be used to pull together research data and improve our understanding of the factors affecting pest outbreaks and pest damage. This can help to identify research priorities by highlighting the key information that needs to be collected to improve this understanding.

Improving Research and Development strategies – Models can be used to explore the implications of different R & D options in assessing which is the best one to follow.

Providing a basis for pest management advice – Models can be used to predict pest outbreaks and the performance of control measures, and so be used as a basis for making recommendations. In the context of training, models can provide a means of simulating the pest problems that farmers face and so give users 'experience' of control.

Let us look in more detail at each of these three attributes, and at specific examples that illustrate the particular value of simulation modelling.

Understanding Pest Population Dynamics and Damage

The first example concerns the model introduced earlier, describing the population dynamics of the brown planthopper (BPH) in tropical rice. The second example concerns a crop–pest model, and how it can provide insights into the factors affecting crop damage.

Brown planthopper in the tropics

The initial objective of the modelling exercise was to account for the dynamics of BPH populations, as observed in historical data. However, before attempting to construct a simulation model, a primary decision tool, an interaction matrix, was used to describe relevant components and relationships in the rice agroecosystem (Fig. 8.4). This description was much wider in scope than the intended simulation model. The purpose was to specify interactions in the wider system portrayed in the interaction matrix, to ensure that key interactions were included in the simulation. Both dots and squares (Fig. 8.4) represent interactions between system components. Squares represent those interactions which were judged to be important to include in the simulation.

A second primary tool, a relational diagram, was also used in conjunction with the interaction matrix (Fig. 8.5). With its origins in industrial systems analysis, the main components in a relational diagram are described by state variables. The value of state variables, such as eggs, nymphs and adults, characterize the system at any point in time. States change

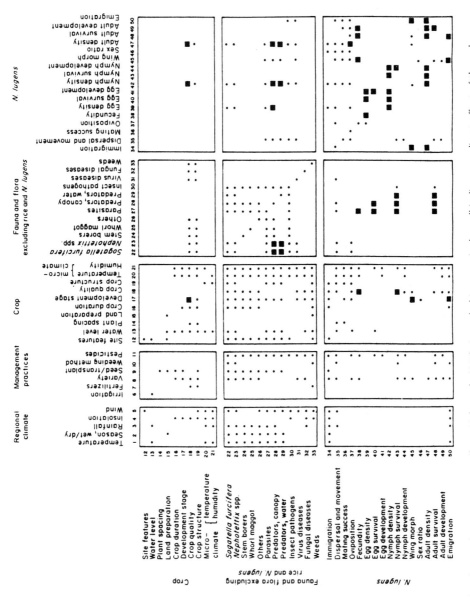

Fig. 8.4. An interaction matrix for the BPH/rice crop system (from Holt et al., 1987). A symbol in a particular cell indicates a direct effect of the column component on the row component. See text for further explanation.

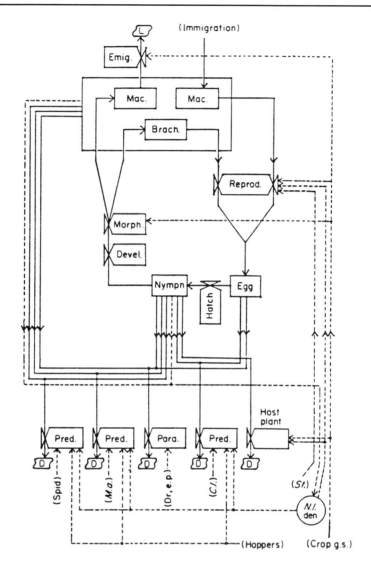

Fig. 8.5. A relational diagram of the *N. lugens* (BPH) simulation model (from Holt *et al.*, 1987). Rectangles denote state variables, circles = auxiliary variables, brackets = driving variables, sinks (⌷) = end points, and valve symbols (☒▭) = factors affecting rates of flow. Key: —— a flow of matter; – – – a link denoting an effect; emig, emigration rate; Morph, morph determination; Devel., development rate; Hatch, hatching rate; Reprod., reproductive rate; Pred., predation rate; Para., rate of parasitism; Host plant, rate of mortality due to host plant; Mac., macropters; Brach., brachypters; Spid., spiders; M.a., *M. atrolineata*; Dr, Dryinidae; e.p., egg parasites; *C.l., C. lividipennis*; *St.*, *Strepsiptera*; *N.l.* den., *N. lugens* density; Hoppers, hoppers other than *N. lugens*; Crop g.s., growth stage of the crop; L, leaving the system; D, death.

according to rate variables. Rates may be constant, for example hatching rate (Fig. 8.5), or be functions of other variables, for example reproductive rate (Fig. 8.5). Driving variables, such as immigrant density and crop growth stage, are so called because they drive the system but are not themselves affected by interactions within the system. For computation in the simulation, the relational diagram is represented by a series of equations that describe state variable changes from one time step to the next, as described earlier.

The initial simulation model of BPH (Holt *et al.*, 1987) included specific mortalities caused by a range of predators and parasites, and included their densities (as well as those of other hopper species) as driving variables. The purpose was to account for the role of different mortality agents in BPH dynamics. Although experimental data are available to describe predation rates individually, a large number of assumptions had to be made to handle the complexity of field interactions in a multi-prey, multi-predator system.

The second objective of the modelling work was to investigate the causes of BPH outbreaks in tropical rice systems. To achieve this, the description of mortality in the model had to be simplified, both because available data did not justify the current level of complexity, and because predator densities could not be driving variables in a fully dynamic version of the model. The resulting, simplified model (Holt *et al.*, 1989) describes mortality due to natural enemies by a density-dependent function (Fig. 8.6). This incorporates the idea of a refuge from predation at low densities, very high mortality due to predators at medium densities, and allows the population to escape from natural enemy action at higher densities. This description of mortality represents a step towards a simpler, more empirical, treatment that has allowed us to

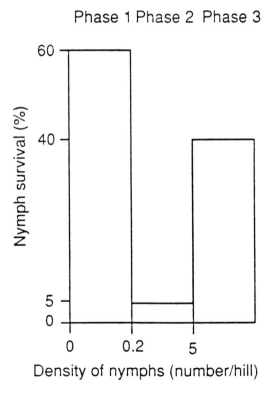

Fig. 8.6. The hypothesis of density-dependent nymphal survival represents a refuge from predation (Phase 1), aggregation of predators to BPH patches (Phase 2) and 'escape' of the BPH population from natural enemy regulatory mechanisms (Phase 3) (from Holt *et al.*, 1989).

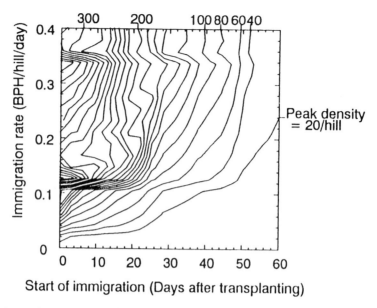

Fig. 8.7. Peak BPH densities resulting from different immigration situations (tropical model simulation results) (from Holt *et al.*, 1989).

explore a range of situations, providing new insights into our understanding of BPH population dynamics.

To illustrate this point, let us consider two outputs from the simulation model. The first output, shown in Fig. 8.7, is a population surface, where the contours represent the peak number of BPH in a crop season as a function of the level and starting time of immigration. This shows that if immigration does not start until 40 days after transplanting, the level of immigration will have to be very high to produce a peak population above 100 BPH per hill, the nominal threshold of concern. If immigration starts much earlier, even low numbers of immigrants per day can cause a problem, if they continue to invade the crop.

A second illustration of the increased understanding that can be gained from simulation models involves the tropical BPH model, described above, as well as a similar model, which includes temperature as a driving variable, for BPH in temperate conditions in China. Examination of the output of these two models has produced a conceptual model that illustrates the behaviour of simulated BPH populations in rice crops, and provides an explicit hypothesis for real populations (Fig. 8.8). A BPH population in Fig. 8.8 can be thought of as a ball on a surface – the ball always tends to roll down-hill. This conceptual model depicts the inherent instability of temperate populations and their tendency to increase (Kuno, 1979), and the stability of tropical populations (Cook and Perfect, 1989) and how this stability can be lost, by high immigration (Loevinsohn, 1984) or by reducing natural enemies by spraying (Kenmore, 1980).

Crop–pest simulation

Where the damage relationship is complicated or interactions with other factors (e.g.

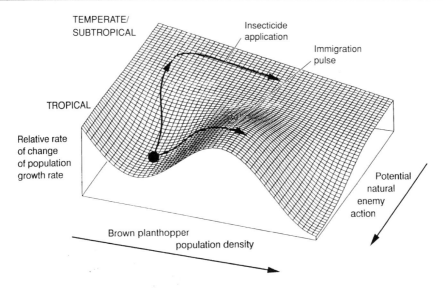

Fig. 8.8. Synoptic model of BPH population change depicting the stability of tropical populations and how that stability might be lost due to insecticide use or exceptional levels of immigration (from Cheng *et al.*, 1993).

drought stress) may interact with pest damage, a simulation of the crop coupled to a simulation model of a pest may be a useful approach (Boote *et al.*, 1983; Rabbinge *et al.*, 1989). SUCROS (simple & universal crop growth simulator) is an example. The crop simulation model is shown in the form of a relational diagram. Pests may have an impact on the crop model at a number of points, depending upon the type of damage caused. The example (Fig. 8.9) is appropriate for an aphid damaging a cereal crop (W. Rossing, pers. comm.). Aphid feeding reduces plant assimilates. Feeding activity also produces honeydew and affects the nitrogen balance in the plant. This in turn has effects on photosynthesis and respiration (maintenance).

An alternative to the relational diagram representation in multi-trophic level models is the 'supply–demand' concept. Dynamics at each trophic level are controlled by the balance of supply and demand for resource (Getz and Gutierrez, 1982). A model of pink bollworm on cotton in Egypt uses this approach (D. Russell, pers. comm.). A certain mass of plant material (leaf, etc.) imposes an assimilate demand for maintenance and growth. Assimilate is supplied through photosynthesis. Pests may impose extra assimilate demands or affect the mass of plant parts. Assuming that the assumptions of the supply–demand process are correct, it is possible to model such effects as compensation for pest damage, impact of irrigation regimes on pest damage, etc., in a relatively mechanistic way. The drawbacks are those referred to in Chapter 6 for wide scope/high detail models: they are difficult to parameterize, difficult to validate and expensive on resources.

Improving Research and Development Strategies

We have already seen above how improved understanding obtained from simulation

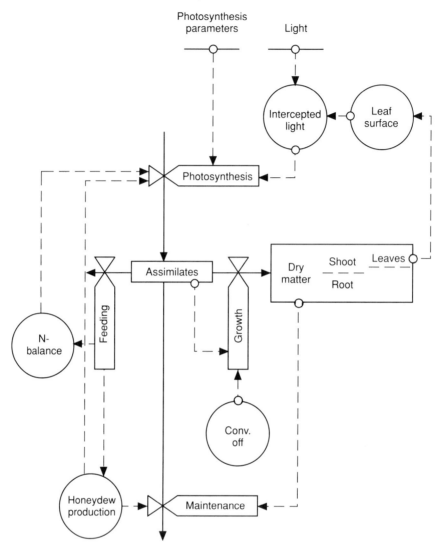

Fig. 8.9. Relational diagram of SUCROS, adapted for injury by cereal aphids (from W Rossing, unpublished course notes).

modelling can help to determine key information gaps to which research might be directed. Here, we focus on more practical aspects of R & D, concerned with the development of pest management options.

Biological control

In Chapter 16, Jeff Waage and Nigel Barlow describe how a population model has been used to investigate the impact that two biological control agents might have on the mango mealybug, an introduced pest first discovered in Ghana and Togo around 1981. The idea was to use this model to determine which of the two bio-control candidates should be released, on the basis of their predicted efficacy in controlling the mealybug. Due to the urgency of the situation, one species was released before the model was complete. However, subsequent work indicates that the model correctly identified the 'best' agent (fortunately, the one released) and provides good prospects for using models in this way for future R & D decisions in biological control.

Pesticide formulation

A second example of R & D decision making concerns the pesticide industry. Like biological control, making decisions on which potential pesticides are to be developed further requires a prediction to be made of the field performance of potential products. In this particular case, simulation modelling was used to investigate the likely practical benefits of more persistent formulations of a synthetic pyrethroid for control of *Heliothis* on cotton (Day and Collins, 1992).

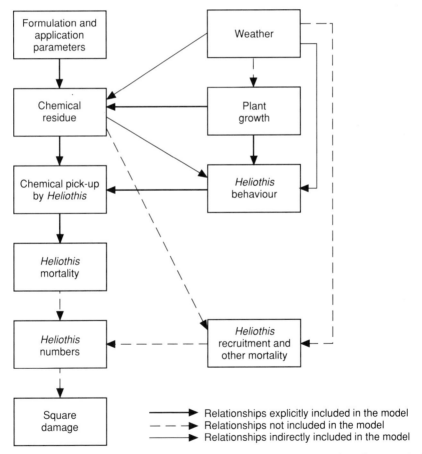

Fig. 8.10. Principal components and pathways describing the effect of a chemical residue on *Heliothis* in cotton (from Day and Collins, 1992). An arrow denotes the relationship 'effects'.

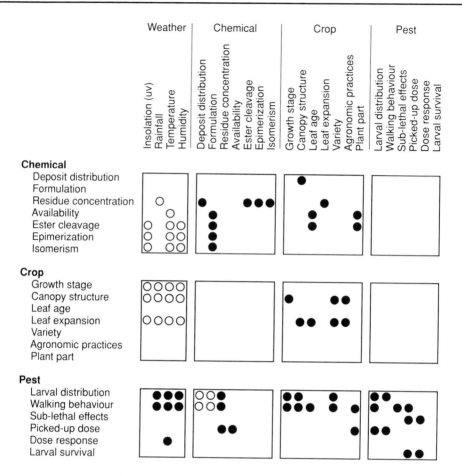

Fig. 8.11. Interaction matrix for first instar *Heliothis* larvae in cotton sprayed with a pyrethroid (from Day and Collins, 1992). Circles indicate that the effect of the column factor on the row factor is known or thought to exist. Solid circles indicate relationships explicitly included in the model.

A model was constructed to describe the mortality of first instar *Heliothis* larvae on cotton during the period following a spray application, based on data gathered from published and unpublished reports. The flow chart in Fig. 8.10 indicates the major components and relationships that need to be considered in assessing the effect on observed damage in the field.

Subsequently, an interaction matrix was constructed to obtain a more detailed description of the system, prior to modelling (Fig. 8.11). Each cell in the matrix shows whether any direct effect of the column factor on the row factor is known or thought to exist. Construction of the matrix was found to be very useful in facilitating the collation of information from different areas of expertise, and to provide a basis for deciding which relationships were directly incorporated in the model. Relationships not explicitly included in the model can still be investigated with simulations, by altering the appropriate parameter values when running the model. For instance, although weather is not directly included in the model, its effect on growth rates in the cotton plant can be investigated by changing growth rate parameters.

A major feature of the problem highlighted by the descriptive phase of analysis was the importance of spatial aspects. A number of components and processes represented in the interaction matrix either include a spatial component, or have different values at different locations within a cotton plant. For example, neither *Heliothis* larvae nor chemical deposits are evenly distributed across the plant surface. Different parts of the plant itself also grow at different rates, affecting growth dilution.

To account for this heterogeneity, a single cotton plant is modelled as a series of discrete patches. Each patch is described by its position in the plant and its surface area. Events relating to the plant surface are modelled separately for each patch. Five types of event are modelled:

- Appearance and expansion of the plant surface.
- Chemical deposition.
- Larval distribution.
- Residue decay.
- Chemical pick-up by the larvae.

Once the model had been constructed, the primary aim of simulation runs was to investigate how the output of the model changed when the two parameters describing a particular formulation of the pyrethroid were altered: namely, the half-life of the biologically active residue, and the toxicity relative to a standard formulation.

Using the best estimates for parameter values, the model predicts first instar mortality remains above 50% for just under 5 days using the current formulation. This performance can be improved by 38% and 57% when both the half-life and relative toxicity are doubled and trebled, respectively. However, if only one of the formulation parameters could be improved, a much greater increase in its value would be required to achieve similar improvements. For example, doubling the half-life alone results in only a 20% increase, while doubling the relative toxicity alone gives an improvement of only a few percentage points. An additional, practical problem is that making the active ingredient more available to the insect also tends to make it more prone to degradation or removal from the plant surface. The net result is that this inverse relationship between the two formulation parameters is likely to severely restrict the potential for improving pesticide performance through reformulation.

Combining this information with an assessment of the feasibility and cost of developing different formulation parameters, ICI could make a decision based on a more rigorous integration of relevant information. The study also highlighted the inadequacy of available data on the behaviour of *Heliothis* on sprayed cotton plants.

Providing a Basis for Pest Management Advice

Given the ability of models to predict, there are a number of distinct ways in which simulation modelling can provide a basis for pest management advice. In particular, they can be employed for the following:

Real-time operation – Given relevant inputs for a specific field at a particular time, real-time simulation can be used to predict the time and/or level of pest attack, and provide advice on tactical, short-term options for that particular case.

Derivation of general rules – Rather than run a model real-time, each time an answer is required, simulation analysis can be used to derive general recommendations, determining which are likely to be the best strategies to adopt for specific sets of conditions.

Let us look at each of these in turn, using several case studies to illustrate.

Real-time operation

When management decisions largely rely on the prediction of future pest attack, the ability of simulation models to predict can make a valuable contribution. In some cases, it is prediction of the timing of events in the population dynamics of a pest which is important. In other cases, it is the prediction of the level of attack which is more important. Let us consider two case studies that illustrate both aspects.

CUTWORMS IN THE UNITED KINGDOM

Cutworms (*Agrotis* spp.) attack valuable vegetable crops in the United Kingdom. In the Thames Valley area, near London, they can cause severe damage to leek crops. The main method of controlling this pest for many years involved the use of insecticide sprays. However, the efficacy of these sprays depended on accurate timing, since spraying too early would not kill cutworm eggs on the leaf, while spraying too late would not kill the fourth instars and above, that have dropped from the plant and feed on the roots below the soil.

To provide growers with advice on the best time to spray, the Agricultural Development and Advisory Service operate a phenological or day-degree model (Bowden *et al.*, 1983). This model, like the well-known PETE (Predictive Extension Timing Estimator) model (Welch *et al.*, 1978), proceeds in daily steps. Each day, the physiological time accrued is calculated, that is, the sum of the day-degrees above and below a species-specific threshold. Each life stage of the pest requires a certain accumulation of physiological time (day-degrees) before development can proceed to the next stage.

Thus, in the case of the cutworm, we know that it takes a certain number of day-degrees above a certain threshold temperature for eggs to mature, hatch, and reach the third instar, the optimal stage at which to spray. On this basis, we can predict the 'best' time to spray by:

1. Estimating the date at which egg laying is at a peak by monitoring moths caught in light or pheromone traps.
2. Monitoring temperature from that date and using the phenological model to determine the number of day-degrees that have been reached to the present time.

With this information, and past temperature records, the number of days it will take to reach the required day-degrees to the third instar can be estimated, and a recommended spray date given (based on expected time at which 80–85% of the larvae have reached the third instar). Table 8.1 shows an example of output from the cutworm model, based on historic temperatures from one site (Taylor, 1984).

During the course of research necessary for developing this predictive model, it was observed that mortality of the younger cutworm instars was closely related to the amount of rainfall: the higher the rainfall, the higher the mortality. With this information, most leek growers in the Thames Valley now use the information provided by the day-degree model to decide the best time to overhead irrigate, when it will cause

Table 8.1. Output from the cutworm simulation model by Taylor (1984) using meteorological data for 1984 from Gatwick airport. The model predicts development and mortality assuming one batch of eggs is laid each night from 31 May. Eggs hatch when the cumulative development is 1.0 greater than on the day they were laid (so the batches laid on 31 May and 1 June both hatch on 16 June, those from 2 and 3 June hatch on 17 June, etc.). Larval development depends on temperature and survival is based on rainfall.

Date	Daily mean temp. (°C)	Rain (mm)	Nightly egg batches		Larval batches		Proportion surviving	
			Cumulative development	Predicted number hatched	Cumulative development	Predicted number expected to reach 3rd instar	For batch to hatch each day[a]	Cumulative survival[b] (=L$_3$ index)
May 31	11.7	2.4	0.04					
June 1	10.7	1.8	0.08					
2	12.1	7.7	0.12					
3	12.8	0.1	0.17					
4	10.7	2.6	0.21					
5	12.3	0.0	0.26					
6	13.8	3.8	0.31					
7	16.1	0.0	0.40					
8	15.8	0.0	0.47					
9	14.9	0.0	0.55					
10	15.3	0.0	0.62					
11	14.0	0.0	0.68					
12	14.2	0.0	0.74					
13	16.3	0.0	0.82					
14	18.2	0.0	0.94					
15	15.0	0.0	1.00					
16	16.0	0.0	1.09	2	0.06			
17	17.2	0.3	1.18	2	0.13			
18	17.8	0.0	1.28	2	0.21			
19	18.4	0.0	1.40	2	0.29			
20	19.8	0.0	1.53	1	0.38			
21	17.9	0.3	1.63	2	0.46			
22	14.5	0.0	1.69	1	0.51			
23	14.5	0.0	1.76	1	0.56			
24	14.6	0.0	1.83	1	0.61			
25	17.1	0.0	1.93	0	0.68			
26	18.4	0.0	2.04	2	0.76			
27	17.3	0.0	2.13	1	0.83			
28	12.4	0.0	2.18	0	0.87			
29	11.7	0.0	2.22	1	0.91			
30	13.4	0.0	2.28	0	0.95			
July 1	12.9	0.0	2.33	1	0.99			
2	13.8	0.0	2.39	0	1.04			
3	13.8	0.0	2.45	1	1.09	2	0.94	1.88
4	14.3	0.0	2.51	0	1.14	2	0.94	3.76
5	16.0	0.0	2.59	1	1.20	0	0.97	3.76
6	17.3	0.0	2.69	1	1.27	2	0.97	5.70
7	17.2	0.0	2.78	2	1.34	2	0.97	7.64
8	21.9	0.0	2.94	2	1.45	1	0.97	8.58
9	16.4	2.5	3.03	0	1.52	3	0.97	11.49
10	18.0	0.0	3.14	2	1.59	1	0.75	12.24
11	16.5	0.1	3.22	2	1.66	1	0.74	12.98
12	16.6	3.3	3.31	1	1.72	0	0.50	12.98
13	16.0	4.5	3.40	2	1.79	2	0.28	13.54
14	16.8	0.0	3.47	1	1.84	1	0.28	13.82
15	15.0	1.4	3.54	1	1.89	0	0.24	13.82
16	15.2	0.0	3.63	1	1.96	1	0.24	14.06
17	16.3	0.0	3.74	1	2.04	1	0.24	14.30
18	18.9	0.0	3.82	1	2.10	1	0.24	14.54
19	15.9	0.0	3.88	0	2.15	0	0.24	14.54
20	13.9	0.0	3.97	1	2.22	1	0.24	14.78
21	16.7	0.0	4.09	1	2.31	1	0.24	15.02
22	19.2	0.0	4.22	1	2.39	2	0.24	15.50
23	19.2	0.0	4.31	2	2.47	2	0.24	15.98
24	17.6	1.3	4.39	0	2.52	0	0.21	16.19
25	15.3	2.4						
Total				**43**		**26**		**16.19**

[a] Daily survival is 1 minus (0.1 × total mm rain during last unit of cumulative larval development).
[b] Cumulative 3rd instar larvae is the product of the number expected to reach 3rd instar × survival value for that day.

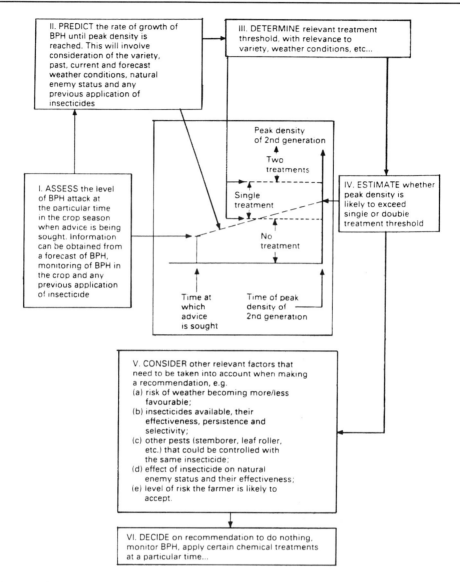

Fig. 8.12. The process of determining recommendations for BPH control (from Holt *et al.*, 1990).

maximum mortality to cutworm larvae. This has replaced the use of pesticides for cut-worm control.

In the example above, information on mortality (caused by rainfall) has been used to enhance the use of a phenological model, based on day-degrees. In other cases, additional information on fecundity and mortality can be entered into the calculation of rate of population development, to predict the magnitude, as well as the timing, of population events. To illustrate this, we turn again to the brown planthopper, this time in China.

BROWN PLANTHOPPER (BPH) IN CHINA – PREDICTIVE MODEL

One of the problems facing rice farmers and their advisers in the Yangtze Delta area of China

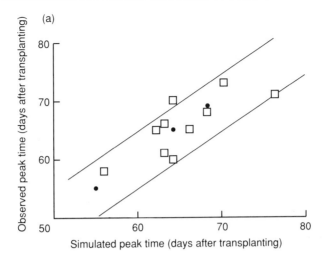

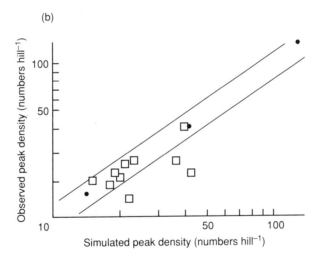

Fig. 8.13. Comparison of (a) time and (b) density of simulated and observed BPH population peaks in ten independent fields (hollow squares) and three fields used to build the model (dots) (from Cheng and Holt, 1990). Lines give an indication of simulated/observed discrepancies: (a) ± 5 days and (b) ± 20%.

is whether to spray against BPH. Since spraying the first generation is generally the most effective way of reducing damage caused by the second generation, a decision, based on a prediction of the size of the second generation, has to be made during the first generation. This problem, as portrayed in Fig. 8.12, is a very common one in pest management.

To predict the size of the second generation, a model was developed for BPH in the Yangtze Delta of China (Cheng and Holt, 1990). The model describes the population dynamics of BPH over three generations (immigration, first and second generation) within a field during the late rice crop. The model is driven by temperature and by numbers of immigrants. Immigration takes place over the first 10–30 days of the cropping season and is assessed by field sampling. This initiates the crop population, the growth of which is determined by temperature conditions over the season.

The model has been 'validated' by comparison with independent population data from ten fields, data from three fields having been used to build the model (Fig. 8.13). Comparing predictions based on the simulation model with those obtained from the regression model, currently used for predicting BPH, the simulation model appears to give a better prediction (Cheng and Holt, 1990).

Derivation of general rules

The case of BPH in China can also be used to illustrate another role that models can play in providing advice, by developing general rules for pest management. In this context, the simulation model is used to carry out a whole series of 'experiments', employing different management strategies and investigating their performance over a wide range of simulated situations. A major goal is to search for 'robust' strategies that perform well, and can be recommended, across a wide range of situations. After considering BPH in China, we then look at general rules for insecticide use and BPH in the tropics and, finally, consider the case of the cattle tick in Australia.

BROWN PLANTHOPPER IN CHINA – RULES FOR SPRAY TIMING

Since field trials indicate a linear relationship between peak density of BPH and yield loss, the performance of different spraying strategies is assessed on the basis of the reduction in peak pest density they cause (Cheng *et al.*, 1990). It is worth noting that a crop model to determine changes in yield loss could be used but, for answering the particular management question posed here, it is thought to have little extra value to add.

The general picture that emerges from the output of a series of simulation runs is:

- If a single application is to be made, 30 days after transplanting (DAT) is the best time to spray. This rule is relatively robust, being true for a range of transplanting times, temperatures, and immigration patterns (Fig. 8.14).
- If more than one spray is to be applied, the first should still be applied at 30 DAT, the timing of the second or subsequent sprays, if needed, having little effect on their effectiveness.

BROWN PLANTHOPPER IN THE TROPICS – RESURGENCE RISK ASSESSMENT

We saw earlier in this chapter, in relation to the conceptual model shown in Fig. 8.8, that the dynamics of BPH in the tropics appears to be very different from that in temperate China. Indeed, we saw that spraying with insecticide can cause resurgence, by reducing the impact of natural enemies on BPH. One may then ask, why spray against BPH at all? There are two answers to this. First, as we saw in Fig. 8.7, high numbers of immigrants can give rise to populations that need to be sprayed, although these immigrants may be the result of insecticide resurgence effects elsewhere. The second situation is that rice farmers may spray to protect their crop from other pests, such as leaf folders, leaf rollers and stem borers. Clearly, what they would want to avoid is, by reducing one pest problem, to increase another, BPH.

To investigate this, the tropical BPH model (Holt *et al.*, 1989), described earlier, has been used to assess the risk of insecticide-induced resurgence of BPH (Holt *et al.*, 1992). An idea of the type of result obtained is shown in Fig. 8.15, which indicates the risk of

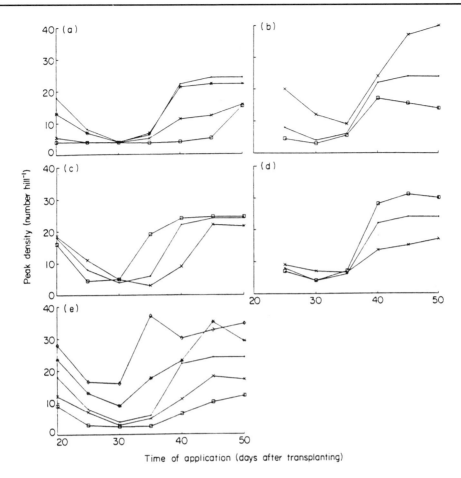

Peak density (number hill⁻¹)

Time of application (days after transplanting)

Fig. 8.14. The effect of application time of a single insecticide treatment on peak density, as modified by the following variables. (a) Type of insecticide: MTMC (half life = 1 day (—); diazinon (3 days)(×); carbaryl (10 days)(*); and carbosulfan (18 days) (□). (b)Temperature: average temperature for August and September (—); favourable to BPH – August temperature 1°C below average, September temperature 1°C above average (×); unfavourable to BPH – August temperature 1°C above average, September temperature 1°C below average (□). (c)Time at which immigrants first enter the crop: -5 DAT (□), 1 DAT (—), and 5 DAT (×). (d) Immigration pattern: concentrated (□), standard (—), and protracted (×) (as used and described in Cheng and Holt (1990) Fig 5). (e) Transplanting time: 22 July (◇), 27 July (*),1 August (—), 6 August (×), 11 August (□). (From Cheng *et al.,* 1990.)

BPH resurgence following the application of an insecticide which kills BPH and natural enemies in equal proportions.

The highest risk of resurgence occurs when the insecticide is applied between 20 and 30 days after transplanting. The degree of risk critically depends on the period it takes for natural enemy populations to recover following the spray. If the area sprayed is small and natural enemies move in quickly from surrounding, unsprayed rice crops, giving 50% recovery within a week or so, the resurgence risk appears negligible, whatever the spray time. These results need to be treated with some caution, until further field testing has been carried out. In the meantime, the type of simulation

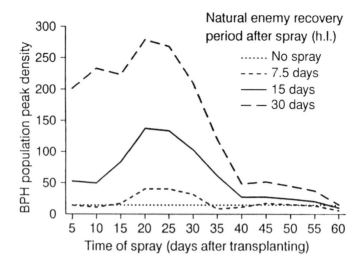

Fig. 8.15. Insecticide induced resurgence showing sensitivity to natural enemy recovery (modified from Holt et al., 1992).

result shown in Fig. 8.15 provides a working hypothesis for making recommendations.

The third case study below again illustrates the value of using simulation models for developing general recommendations but, in this case, for a pest which is endogenous, and where the cost of carrying out field experiments to assess management strategies is extremely expensive.

CATTLE TICK

The cattle tick, *Boophilus microplus*, is a problem in Australia, Brazil, and other areas, not only as a pest in its own right but also as a vector of disease. Recommendations are required on a number of possible options that cattle managers might take in adapting to cattle tick problems. In Australia, where producers vaccinate their cattle against tick-borne diseases, the concern is with the control of ticks as a pest themselves.

On the basis of extensive research at the CSIRO Division of Entomology in Australia, a series of models have been developed to provide support for both research and extension decision making. The earliest modelling work was aimed at pulling together the existing ecological knowledge on the tick to help understand the key determinants of its population growth (Sutherst and Dallwitz, 1979).

From this initial study, a simpler population model was constructed, that allowed various management options to be assessed and so provided a basis for recommending tick control strategies. This simulation model allows changes in climatic conditions, breed of cattle, stocking density, and dipping practice, amongst others. An outline of the structure of the model can be found in the interaction matrix described in Chapter 2. Details of the work can be found in Sutherst *et al.* (1979) and Norton *et al.* (1983, 1984).

By changing appropriate variables, simulation runs of the model have been used to assess the impact of different integrated pest management options on tick population

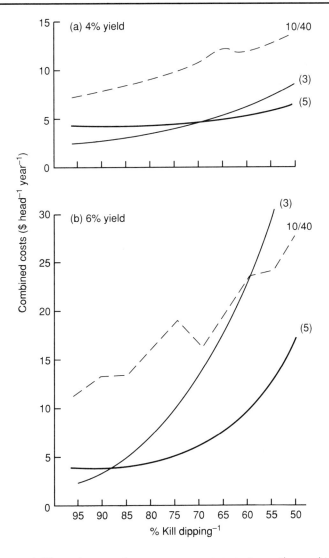

Fig. 8.16. The impact of different integrated pest management strategies on the combined costs of tick damage and control (simulation). Resistance level of the cattle is expressed as the percentage yield of ticks – the lower the percentage yield, the higher the resistance level. Two strategic dipping strategies are assessed: 3-dippings per year and 5-dippings per year (both at three week intervals). Different percentage-kills can be associated with method of acaricide application (dipping, hand-spraying, etc.), acaricide dose, and percentage of cattle mustered and treated.

dynamics and tick damage. Used in this way, the cattle tick simulation model has given answers to specific management questions (Fig. 8.16).

Rather than attempting to predict tick populations at a particular time in a particular place, the approach adopted was to start with control measures, and to use the tick model to search for 'robust' control strategies, defined as those strategies that operate satisfactorily over a wide area of parameter space. By adopting this approach, the problem of applying the model to a particular situation, in a predictive sense, is reduced to the much simpler task of assessing whether the parameter and variable values for that situation are likely to lie within certain bounds, namely, those which define situations in which the strategy would be acceptable to the producer.

Used in this way, the results of the cattle tick simulation model have been extremely valuable in terms of practical implementation. Two specific examples are given below:

- In the late 1970s and early 1980s there was rapid adoption of Zebu-cross cattle in Queensland, in the face of increasing acaricide resistance. Since there had been little field experimentation with these animals, it was not clear to what degree the recommended dipping strategies for European (susceptible) breeds could be modified. By including Zebu-cross animals in the model, and trying out numerous dipping strategies, simulation analysis provided 'instant experience'.
- By simulating the practice of pasture spelling, which involves removing the cattle from the main paddock for a period of 4 weeks or so, it was shown that this can be an extremely effective means of controlling tick populations in Zebu-cross herds, without resort to acaricides. Previous field trials, using susceptible European breeds, had found that pasture spelling was not a promising means of control.

The results of this modelling work have also provided a framework for making general recommendations based on the two key variables, climatic favourability and host resistance (Fig. 8.17).

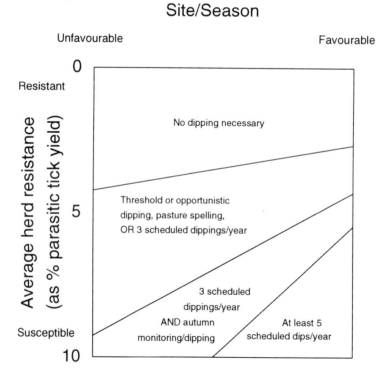

Fig. 8.17. General recommendations for tick control based on climatic favourability and host resistance (after Norton *et al.*, 1983).

Tick species

		One major tick species	More than one major tick species
One major host	Tick fever 'controlled'	Australia	
	Tick fever 'not controlled'	Brazil	
More than one major host	Tick fever 'controlled'		
	Tick fever 'not controlled		Many countries in Africa

Fig. 8.18. Broad classification of tick problems on the basis of biological/ecological differences.

WIDER VALUE OF THE TICK MODEL

This particular example of the use of simulation modelling to answer a variety of biological and management questions can also be used to illustrate how simulation modelling can have much wider application. The extensive work carried out in Australia on tick ecology, tick control methods, and modelling, can be of value in tackling *Boophilus* and other tick problems throughout the tropics and subtropics: in particular, efforts are being made to link the Australian work with that of several groups in Africa (Sutherst, 1987). Apart from general ecological lessons that have been learned about the key factors affecting population dynamics, and the importance of monitoring engorging ticks on the animal, the modelling work provides a particularly important means of extrapolating from the Australian experience to problems in other countries.

This is particularly apparent with the climatic matching program, CLIMEX (described in Chapter 6), which enables a first-shot to be made at estimating whether certain species of ticks are likely to survive at a particular location and, if so, the growth potential of these ticks. For *Boophilus microplus*, CLIMEX can be used in combination with the management model to make preliminary estimates of the performance of tick control strategies.

Clearly, in attempting to 'transfer' experience and technology from Australia to other countries, it is crucial that all of the differences in the respective problems in each country are taken into account. Apart from political and socioeconomic differences, there are major differences in the biological and ecological nature of the problem, as summarized in Fig. 8.18. This again highlights the need for clearly determining the precise dimensions of pest problems in designing research and implementation strategies. It was with this major objective in mind that a series of workshops were carried out in Brazil in 1988. Employing a range of primary decision tools, the workshops aimed to produce guidelines for future research and implementation needs (Norton and Evans, 1989).

WHAT DECISIONS NEED TO BE MADE ABOUT THE TYPE OF SIMULATION MODEL TO BUILD?

Having seen what simulation models are, and considered how they can be useful for a range of decision-making problems related to pest management, we now turn our attention to the decision problems involved in developing a simulation model. Are there any clear guidelines we can give on how to do it?

In the previous chapter, the importance of specification was emphasized. This involves two steps, problem and model specification.

Problem Specification

There are three key processes that need to be considered in problem specification:

Identify the information need – This arises from the decision problems that pest managers, policy makers, industry, or research and extension scientists face. The approach and techniques described in Chapters 1 to 5 provide many examples of how this might be achieved.

Clarify the precise questions that are being asked – What level of resolution is required in order to answer these questions? This is often not as easy as it may seem, and again, as above, requires close collaboration with the decision maker. The most efficient model is that which achieves the required level of resolution with the minimum of effort.

Determine the key components and relationships – These need to be incorporated in the model in order to answer the questions posed.

Model Specification

This involves examining the modelling options available and choosing the technique and design of simulation model most appropriate for the scope and detail of modelling required, as determined from problem specification (cf. Fig. 6.4).

Scope of the model – In some cases, simulating the population dynamics of a single pest species may be the most appropriate thing to do. In other cases, it may be more appropriate to construct a series of linked simulation models, of the crop, several pest species and several natural enemy species. Even single-species models may represent wider systems by inclusion of other factors as variables but, if interactions are complex, difficulties may arise.

Detail of the model – The dynamics of a single species may be described by few or many elements and interactions, as we have seen in this chapter. The rate of population increase may be described by a single parameter 'r', as in the apple mildew example, or by a whole series of parameters representing such processes as fecundity, survival and development, as described for the rice brown planthopper.

The examples given in this chapter will hopefully provide some clues on how to choose the most appropriate scope and detail of particular simulation models. To be more precise than this would be difficult, since a major point we have been making is that each problem must be considered independently, the most appropriate model structure being determined by the nature of the problem and the questions being asked. There is one general rule that we would recommend, however, and that is to keep the model as simple as is needed to do the job.

REFERENCES

Barlow, N.D., Norton, G.A. and Conway, G.R. (1979) A systems analysis approach to orchard pest management. Mimeographed report, Environmental Management Unit, Department of Zoology and Applied Entomology, Imperial College, Ascot, UK.

Boote, K.J., Jones, J. W., Mishoe, J.W. and Berger, R.D. (1983) Coupling pests to crop growth simulators to predict yield reductions. *Phytopathology* 73, 1581–1587.

Bowden, J., Cochrane, J., Emmett, B.J., Minall, T.E. and Sherlock, P.L. (1983) A survey of cutworm attack in England and Wales and a descriptive population model of *Agrotis segetum*. *Annals of Applied Biology* 102, 29–47.

Cheng, J.A. and Holt, J. (1990) A systems analysis approach to brown planthopper control on rice in Zhejiang Province, China. I. Simulation of outbreaks. *Journal of Applied Ecology* 27, 85–99.

Cheng, J.A., Norton, G.A. and Holt, J. (1990) A systems analysis approach to brown planthopper control on rice in Zhejiang Province, China. II. Investigation of control strategies. *Journal of Applied Ecology* 27, 100–112.

Cheng, J.A., Holt, J. and Norton, G.A. (1993) A systems approach to planthopper population dynamics and its contribution to the definition of pest management options. In: Denno, R.F. and Perfect, T.J. (eds) *The Ecology and Management of Planthoppers*. Chapman and Hall, New York. (in press)

Cook, A.G. and Perfect, T.J. (1989) The population characteristics of brown planthopper, *Nilaparvata lugens*, in the tropics. *Ecological Entomology* 14, 1–9.

Day, R.K. and Collins, M.D. (1992) Simulation modeling to assess the potential value of formulation development of lambda-cyhalothrin. *Pesticide Science* 15, 45–61.

de Wit, C.T. and Goudriaan, J. (1978) *Simulation of Ecological Processes*. Pudoc, Wageningen, The Netherlands.

Getz, W.M. and Gutierrez, A.P. (1982) A perspective on systems analysis in crop production and insect pest management. *Annual Review of Entomology* 27, 447-466.

Holt, J., Cook, A.G., Perfect, T.J. and Norton, G.A. (1987) Simulation analysis of brown planthopper population dynamics on rice in the Philippines. *Journal of Applied Ecology* 24, 87–102.

Holt, J., Wareing, D.R., Norton, G.A. and Cook, A.G. (1989) A simulation of the impact of immigration on brown planthopper population dynamics in tropical rice. *Journal of Plant Protection in the Tropics* 6, 173–187.

Holt, J., Cheng, J.A. and Norton, G.A. (1990) A systems analysis approach to brown planthopper control on rice in Zhejiang province, China. III. An expert system for making recommendations. *Journal of Applied Ecology* 27, 113–122.

Holt, J., Wareing, D.R. and Norton, G.A. (1992) Strategies of insecticide use to avoid resurgence of *Nilaparvata lugens* (Homoptera: Delphacidae) in tropical rice: A simulation analysis. *Journal of Economic Entomology* 85, 1979–1989.

Kenmore, P.E. (1980) Ecology and outbreaks of a tropical insect pest of the green revolution, the brown planthopper, *Nilaparvata lugens* Stal. Unpublished PhD thesis, University of California, Berkeley.

Kuno, E. (1979) Ecology of brown planthopper in temperate regions. In: *Brown Planthopper Threat to Rice Production in Asia*. International Rice Research Institute, Los Banos, Philippines.

Loevinsohn, M.E. (1984) The ecology and control of rice pests. Unpublished PhD thesis, University of London.

Norton, G.A. and Evans, D.E. (1989) *Report on a Series of Workshops on Tick and Tick Borne Disease Problems, Held in Brazil 1–22 December, 1988*. EMBRAPA, Brasilia.

Norton, G.A., Sutherst, R.W. and Maywald, G.F. (1983) A framework for integrating control methods against the cattle tick, *Boophilus microplus*, in Australia. *Journal of Applied Ecology* 20, 489–505.

Norton, G.A., Sutherst, R.W. and Maywald, G.F. (1984) Implementation models: the case of the Australian cattle tick. In: Conway, G.R. (ed.) *Pest and Pathogen Control: Strategic, Tactical, and Policy Models*. John Wiley and Sons – for the International Institute for Applied Systems Analysis, pp. 381–394.

Rabbinge, R., Ward, S.A. and van Laar, H.H. (eds) (1989). *Simulation and Systems Management in Crop Protection.* Pudoc, Wageningen, The Netherlands.

Sutherst, R.W. (ed.) (1987) *Ticks and Tick Borne Diseases.* ACIAR Proceedings No.17, ACIAR, Canberra.

Sutherst, R.W. and Dallwitz, M.J. (1979) Progress in the development of a population model for the cattle tick, *Boophilus microplus. Proceedings of the 4th International Congress in Acarology, 1974,* pp. 557–563.

Sutherst, R.W. and Maywald, G.F. (1985) A computerised system for matching climates in ecology. *Agriculture, Ecosystems and Environment* 13, 281–299.

Sutherst, R.W., Norton, G.A., Barlow, N.D., Conway, G.R., Birley, M. and Comins, H.N. (1979) An analysis of management strategies for cattle tick (*Boophilus microplus*) control in Australia. *Journal of Applied Ecology* 16, 359–382.

Taylor, C.G. (1984) An investigation of a predictive system developed to improve timing of control measures against cutworms in vegetables. Unpublished MSc thesis, University of London.

Welch, S.M., Croft, B.A., Brunner, J.F. and Michels, M.F. (1978) PETE: An extension phenology modeling system for management of multi-species pest complex. *Environmental Entomology* 10, 425–432.

9

Rule-based Models

J. Holt[1] and R.K. Day[2]

[1]*Natural Resources Institute, Central Avenue, Chatham Maritime, Chatham ME4 4TB, UK:* [2]*International Institute of Biological Control, PO Box 76250, Nairobi, Kenya*

Introduction

Rule-based modelling is the qualitative equivalent of conventional quantitative simulation modelling. Without the need to represent relationships between components of a system using mathematical equations, a rule-based approach provides a means by which subjective knowledge about a system can be used to build models.

Both conventional simulation models and rule-based models simulate how system components respond over time to changes in other components. In conventional models, the components are represented by real number variables and changes in variables are described by equations. In rule-based models, components are represented by a small set of discrete states and changes are described by 'if–then' type rules.

Many expert systems use if–then type rules as their basic building block (Holt *et al.*, 1990; Jones *et al.*, 1990). In this respect, rule-based models and expert systems are very similar; expert systems are discussed in Chapter 11. In this chapter, however, we reserve the term rule-based model for the use of rules to simulate the dynamics of systems over time.

Rule-based models can be built in any computer language which incorporates logical operators (and, or, greater than, etc.), if–then statements and the capacity for iteration. This includes all conventional programming languages (Basic, Pascal, C, etc.), databases (dBASE, FoxPro, etc.), spreadsheets (Supercalc, Lotus 123, etc.), and some expert system shells. Many expert systems shells are not designed primarily for model building and it is sometimes awkward to instruct the software to iterate (that is, to re-apply the same set of rules for each step of a time sequence). An example of a simple rule-based model built using a spreadsheet is given in Chapter 10.

Role of Rule-based Models

Rule-based models can use information at lower levels of resolution than conventional models. The lowest level of resolution is simply knowing that a particular model component would increase, decrease or stay the same, in response to other specific changes. More usually five or six states of each component are defined (Starfield and Bleloch, 1986), allowing some measure of the extent of increase or decrease to be incorporated, for instance 'increase' or 'large increase'. Experience with rule-based models is showing that this number of states is all that is required for the analysis of a range of environmental problems (I.R. Noble, pers. comm.). In most cases it is also unreasonable to expect that consistent qualitative judgements can be made at levels of resolution much greater than this.

Incomplete 'current state' knowledge can be incorporated (and therefore its implications tested) in an explicit way using the rule-based format. As with expert systems, updating the model is relatively simple, if improved information becomes available.

A rule may be used to capture knowledge about a process without introducing additional assumptions, frequently unavoidable in mathematical representations of the same basic idea. For example, consider a hypothetical relationship between plant density and seed production per plant. The relationship is understood qualitatively, as follows. At very low densities, seed production per plant is low because this species tends to be shaded by taller species. The effects of inter-specific competition decline with increasing plant density, until at high densities, seed production per plant reaches a maximum. At very high densities, however, intra-specific competition reduces seed production. This can be represented by the following set of rules:

IF   plant density is very low   THEN   seed production per plant is low
IF   plant density is low   THEN   seed production per plant is medium
IF   plant density is medium   THEN   seed production per plant is high
IF   plant density is high   THEN   seed production per plant is very high
IF   plant density is very high   THEN   seed production per plant is medium

In the example, the real number equivalents of 'low', etc. are not known, hence the use of such linguistic values. The important point is that 'low plant density' has a consistent meaning throughout the model.

An equivalent quantitative representation might be an 'empirical' curvilinear function with a maximum at high plant density. Extra assumptions are unavoidably added through the details of the shape of the function and the need to guess numerical values. The underlying knowledge can then be difficult to separate from the mathematical representation.

The above example illustrates one of the major advantages of rule-based models, namely that rules are directly comprehensible. The rule-based format often corresponds to a manager's state of knowledge of a system, so that managers and biologists can themselves structure and evaluate models. For this reason, rule-based models have particular value as inter-disciplinary tools.

The way in which a system is described by a model determines the type of conclusions that we might draw from the modelling exercise. Rule-based models have some affinities with analytical models in that they are useful for looking at the principles which govern the behaviour of systems as well as the relative impact of different components in a system. In contrast, conventional quantitative simulation models are often aimed at more absolute evaluation or prediction. If some of the relationships in a system cannot be precisely quantified, however, conventional simulation models tend to be restricted to conclusions in the same sphere as rule-based models.

The tendency with conventional simulation models is to include detail where detailed knowledge is available, and use judgement and estimate where it isn't. The format of rule-based models (with just a few states of *all* components), discourages such imbalanced inclusion of detail. Because detail is kept in check by self-imposed constraints, rule-based models are particularly useful when a model must incorporate a large number of components (models wide in scope and shallow in detail, Chapter 6). In such cases it is probably unlikely that enough of the interactions between system components would be quantified to an extent which would justify conventional simulation.

Rule-based models are not likely to be an appropriate choice when the absolute amount of a component or the precise details of changes in particular components are pertinent to the questions being asked with the model (Knight and Mumford, 1992).

There seem therefore to be two types of situation in which rule-based models, rather than conventional models, may be employed: when the level of description of the system provided by a rule-based model is sufficient to analyse the problem in question, and when a quantitative model would be useful but the information to build it is not available *and* the problem is such that some useful conclusions may be drawn from a rule-based approach.

Having attempted to define a niche for a rule-based modelling approach, it should be said that the line between conventional quantitative simulation models and rule-based models is not distinct. Most conventional simulation models contain 'if–then' operations of some form and it is only the recent interest in applications of expert system techniques of knowledge representation which has crystallized the term 'rule-based model'. Conventional and rule-based modelling approaches can be combined; the resulting mixtures have been called 'rule-enhanced' models (Starfield and Bleloch, 1986).

To illustrate how pest systems might be represented by rule-based models, two examples are described. The first looks at the problem of predicting termite attack in crops grown in a seasonally flooded coastal delta in Sudan – the model simulates the dynamics of termite populations and their foraging behaviour. The second looks at the problem of forecasting armyworm outbreaks in East Africa – this model simulates the impact of meteorological and other components on moth movement and population processes.

CASE STUDY – TERMITES IN SUDAN

The problem of termite attack on crops in Sudan presents a useful case study for a rule-based model. It was developed with the help of M. J. Pearce, of the Natural Resources Institute. Seasonal flooding of a delta area on the Red Sea coast of Sudan allows crops to be grown in those areas which receive flood water. With a first flood in July/August and a second flood in September/October, any location can receive none, one or two floodings. The objective was to provide some evaluation of the risk of termite damage to crops in different locations and seasons. To do this a rule-based model of the termite population at a location was constructed. Each model component had two or more states:

Month:	Jul/Aug, Sep/Oct, Nov/Dec, Jan/Feb, Mar/Apr, May/Jun
Soil:	Saline, Stony, Sand, Clay
First flood:	Yes, No
Second flood:	Yes, No
Crop residues:	Yes, No
Shrubs:	Yes, No
Rain:	Yes, No

The termite components were:

Termite population size: 0 − 10 (ordinal measure)
Change in size: −2, −1, 0, +1, +2 (corresponding to large decrease, decrease, same, increase, large increase)
Immigration: 0 − 5 (same measure as population size)
Termite foraging: 0, 1, 2, 3, 4 (none, slight, moderate, high, very high)

After detailed consideration of the processes affecting termites in each two-month period, a series of rules were devised to reflect the resulting impact of the various possible situations on termite population change and foraging behaviour, such as:

IF   month is Jul/Aug
and   soil is sand
and   first flood covered area
and   shrubs are present or crop residues are present

THEN   foraging is slight (state 1)
and   population change is decrease (−1).

The complete rule set is summarized in Table 9.1. The third and fourth rules from Table 9.1 are written in full (and combined as one rule) above. A brief description of some of the key processes associated with the November/December period might help to illustrate how the rules were devised. There are 14 rules for this period.

Rules 1 and 2 − sand soils, flooded and some rain − leads to large increase in termite population size (+2) and high foraging activity (3). Floods mean that crops will be planted, so there is a good food supply. Light rain at this time helps foraging activity.

Rules 3 and 4 − Lack of rain means the soil is unsuitable for 'gallery' construction and foraging is impaired.

Rules 5 and 6 − Despite failure to flood this year, the presence of shrubs indicates an accessible water table and termites are able to persist.

Rules 7 and 8 − Sand soils, no flooding and no available water table, leads to a large decrease in the termite population.

The rules for clay soils are similar, except that termites can persist under the harshest conditions (Rule 14) because a water table is always available in these soils.

Termite activity (which is assumed to be some product of population size and foraging activity) can be measured with wooden baits. Simulations with the prototype model were compared with field data. This raised several important points:

- Our assumptions about the persistence of termites in clay soils, or when an available water table is present, need to be examined. In the 1985/1986 season, following an absence of flooding in the 1984/1985 season, the modelled population reached very high levels but observed termite activity was actually low (Fig. 9.1a). If instead, we assume decreases in population size take place in a manner equivalent to sandy soils, without an available water table, model predictions are improved (Fig. 9.1b).
- Increases and decreases in activity were much more rapid than suggested by the model. It may be either that population density change is more rapid than was

Table 9.1. Rule-based termite population model – population change and foraging activity predicted for different states. Information not required if state unspecified. (Rules devised with the help of M.J. Pearce, NRI.)

Month	Soil	Flood 1	Shrubs	Crop residues	Flood 2	Rain	Change	Foraging
	stony						−2	0
	saline						−2	0
J/A	sand	y	y				−1	1
J/A	sand	y		y			−1	1
J/A	sand	y	n	n			−2	0
J/A	sand	n	y				0	1
J/A	sand	n		y			0	1
J/A	sand	n	n	n			−2	0
J/A	clay	y	y				−1	1
J/A	clay	y		y			−1	1
J/A	clay	y	n	n			−2	0
J/A	clay	n	y				0	1
J/A	clay	n		y			0	1
J/A	clay	n	n	n			−2	0
S/O	sand	y			y		−2	1
S/O	sand	y			n		0	1
S/O	sand	n	y		y		−1	1
S/O	sand	n	n		y		−2	0
S/O	sand	n	y		n		0	1
S/O	sand	n	n		n		−2	0
S/O	clay	y			y		−2	0
S/O	clay	y			n		0	1
S/O	clay	n			y		−1	1
S/O	clay	n			n		0	1
N/D	sand	y				y	2	3
N/D	sand				y	y	2	3
N/D	sand	y				n	1	1
N/D	sand				y	n	1	1
N/D	sand	n	y		n	y	1	2
N/D	sand	n	y		n	n	0	1
N/D	sand	n	n		n	y	−1	0
N/D	sand	n	n		n	n	−2	0
N/D	clay	y				y	2	3
N/D	clay				y	y	2	3
N/D	clay	y				n	1	1
N/D	clay				y	n	1	1
N/D	clay	n			n	y	1	2
N/D	clay	n			n	n	0	1
J/F	sand	y					1	2
J/F	sand				y		1	3
J/F	sand	n	y		n		0	1
J/F	sand	n	n		n		−2	0
J/F	clay	y					2	3
J/F	clay				y		2	3
J/F	clay	n					0	1
M/A	sand	y					2	4
M/A	sand				y		2	4
M/A	sand	n	y		n		−1	1
M/A	sand	n	n		n		−2	0
M/A	clay	y					2	4
M/A	clay				y		2	4
M/A	clay	n					−1	1
M/J	sand	y					0	1
M/J	sand				y		0	1
M/J	sand	n	y		n		−1	1
M/J	sand	n	n		n		−2	0
M/J	clay	y					0	2
M/J	clay				y		0	2
M/J	clay	n					−1	1

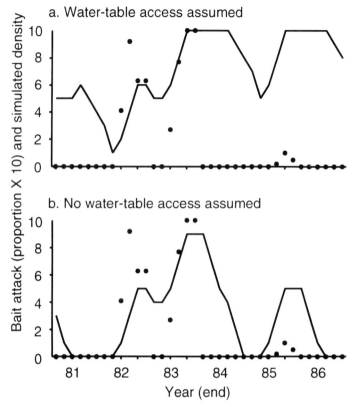

Fig. 9.1. Observed proportion of baits attacked by termites (points) compared to simulated termite density (lines) for two alternative assumptions relating to water availability during the dry season.

believed, or that rapid changes in bait attack are due simply to changes in foraging behaviour.

- The modelled population density was a better predictor of baiting observations than either modelled foraging or modelled activity (a product of density and foraging). This suggests that our assumptions about foraging are incorrect.

CASE STUDY – ARMYWORM IN EAST AFRICA

A rule-based model is appropriate for modelling armyworm populations as there is little quantitative data available but a large amount of experience amongst researchers.

There are two distinct components to the life history of *Spodoptera exempta*:

- Migration – Adult moths can migrate up to several hundred kilometres from their development site, and cause new outbreaks of larvae where they are concentrated by convergent wind flow, usually in the vicinity of storms.
- Breeding – Once the adults have landed, mating and egg laying takes place. A second generation can occur at the same site if conditions are suitable.

Table 9.2. Weather categories in the model.

- Dry: D
  Little or no rain. No convergent windflow

- Light rain: LR
  Widespread rain but no storms. No convergent windflow

- Isolated rainstorms: IS
  Some large storms. Convergent windflow in vicinity of storms

- Occasional widespread rainstorms: OS
  Widespread storms separated in time. Strong convergent windflows

- Frequent widespread rainstorms: FS
  Frequent storms. Some convergent windflows, but due to the high frequency in space and time, the effects on airborne moths assumed to cancel out

The Model

The current model describes the dynamics of armyworm populations for a single unit of space, such as a degree square. A time interval of one week is used, so there are five population age classes; adults and eggs, young larvae, old larvae, prepupae and pupae.

The size of the population in each age class is modelled using a relative scale, as none of the processes in the model is density dependent. A logarithmic scale is used, so that the population size at a particular point on the scale is e^1 times the population at the previous point and e^{-1} times the population at the next point up the scale.

Weather is thought to play a major part in the dynamics of armyworm population development, as well as being important in migration. Five categories of mutually exclusive weather are used, as shown in Table 9.2. These categories were chosen as the minimum that appeared necessary to describe the major effects of weather on armyworm populations.

Female *S. exempta* can lay up to 1000 eggs in the space of a few days and the major limiting factor is the availability of water, though only a small amount is required. Table 9.3 gives the fecundity rates in the model, which are determined by the prevailing weather.

There are thought to be three principal causes of larval mortality; poor food quality, drowning of very young larvae by heavy rain, and natural enemies, particularly pathogens, which are also more prevalent during wet periods.

In non-crop areas, armyworm feeds on grasses and, until the first rains of the season,

Table 9.3. Model fecundity.

Weather	Fecundity[a]	Equivalent eggs per female
D	1	5
LR	4	110
IS	5	297
OS	6	807
FS	6	807

[a]Fecundity value is the number of points the population moves up the logarithmic population scale.

Table 9.4. Larval mortality due to food quality.

Food quality	Larval mortality[a]	Equivalent % mortality
Very low	4	98
Low	3	95
Medium	2	86
High	1	63
Very high	0	0

[a] Mortality value is the number of points the population moves down the logarithmic population scale.

the food quality can be assumed to be very low, if there is any grass present. Once rains have fallen, a flush of young grass rapidly grows, which has a high nitrogen and mineral content, so that within a week there is very high quality food available for any young armyworm larvae present. Thereafter the quality of the grasses declines through the growing season, as the plants age. In the model we make the following assumptions concerning the quality of food available to the armyworm larvae.

1. Food quality is described by a five-point scale (Table 9.4).
2. The flush of vegetation at the start of the growing season does not occur until one of the three categories of storm occurs.
3. Until the time interval of the first storm, food quality is very low.
4. In the interval after the first storm, the food quality rises to very high if the storms were widespread (OS or FS), but only rises to high if the storms were isolated, as in the latter case it can be assumed that many parts of the area will receive insufficient rain to initiate the flush.
5. Food quality declines by one point on the scale every three time intervals, until very low is reached, after which there is no further decline.
6. If the first storms are isolated, the food quality declines as in **5** until the first widespread storms. The following interval the quality rises to very high.

Table 9.4 shows the mortality due to food quality, and the equivalent percentage mortality. This mortality is assumed to occur at the transition from eggs to young larvae as, if food quality is poor, the larvae die during the early instars.

Very young larvae are also susceptible to drowning, which is only likely to occur when heavy rain is widespread. Heavy rain and overcast conditions also increase mortality due to pathogens so, in the model, all egg and larval mortality not due to food quality effects is contained in a single relationship. Table 9.5 shows the model mortality rates due to direct weather effects. This also takes effect at the transition from eggs to young larvae in the model. Minimum mortality is set at 63% (1 on the logarithmic scale) to include all the other egg and larval mortality that occurs due to predation and parasitism, but about which insufficient is known to include any causal relationship in the model.

Armyworm outbreaks usually occur when sufficient individuals have become aggregated and breed synchronously. This is thought to occur when mesoscale convergent windflows concentrate airborne moths. In the model there are three levels of population aggregation, determined by the prevailing weather at the adult/egg stage. Table 9.6

Table 9.5. Larval mortality due to weather.

Weather	Larval mortality[a]	Equivalent % mortality
D	1	63
LR	1	63
IS	1	63
OS	3	95
FS	4	98

[a] Mortality value is the number of points the population moves down the logarithmic population scale.

Table 9.6. Aggregation of airborne moths.

Weather category	Aggregation
D	Low
LR	Low
IS	Medium
OS	High
FS	Low

shows the level of aggregation for the five weather categories. The degree of aggregation for a cohort persists until emergence, when the degree of aggregation for the new generation is again determined by the prevailing weather.

The degree of emigration of adult armyworms from the emergence site is thought to be determined by two main effects. Firstly, if the moths developed from a population of gregarious larvae, they are more likely to be long fliers. Secondly, when rain is present, moths are less likely to emigrate long distances, as they can obtain the moisture required for maturation locally and also the windflow in the vicinity of storms is likely to prevent long distance migration. As we are using a degree square as the unit space, emigration means moths flying out of that degree square, so the levels of emigration used are lower than is thought to occur from a single outbreak site. Table 9.7 shows the model emigration rates.

Immigration is an input in the current model. The net effect of immigration on the population of adults depends on the relative size of the immigrant and resident populations. Table 9.8 shows the rules used.

Results

Few results have been obtained with the model so far. The objective is to validate the model using data from East Africa, and then to run the model on further data to test its performance.

The model will also be used to investigate the combinations of conditions that

Table 9.7. Emigration rate[a].

	Aggregation		
Weather	Low	Med.	High
D	2	4	5
LR	1	3	4
IS	1	3	4
OS	0	1	3
FS	0	1	3

[a] Emigration rate is the number of points the population moves down the logarithmic population scale as a result of emigration. Equivalent emigration rates for model rates of 0, 1, 2, 3, 4 and 5 are 0%, 63%, 86%, 95%, 98% and 99.99%, respectively.

Table 9.8. Immigration.

Resident population	Immigration population	Net population
N	$<N$	N
N	N	$N+1$
$<N$	N	N

Table 9.9. Example output, using real weather data from Kenya, and one pulse of immigration at different times (zero otherwise).

Time interval of immigration	Weather	Population of late instar larvae $(t+2)$	Aggregation
1	D	1	Low
2	D	1	Low
3	D	1	Low
4	LR	4	Low
5	IS	6	Medium
6	OS	8	High
7	OS	7	High
8	FS	10	Low
9	IS	6	Medium
10	OS	6	High

appear most likely to result in outbreaks. Table 9.9 shows a summary of ten runs of the model in which a single pulse of immigration was used at a different time interval each run. The sequence of weather data in this case is real data from Kenya.

Although the highest population of late instar larvae results after immigration at 8 weeks, those larvae show only a low level of aggregation. (Note that they could, however, serve as a source of migrants for outbreaks in the following generation.) With this pattern of weather, immigrants arriving in weeks 6 or 7 are most likely to cause outbreaks, with a fairly high population of highly aggregated larvae resulting.

CONCLUSION

Computation of Population Processes

The two case studies help to illustrate some of the difficulties as well as the benefits of rule-based models. One of the principal difficulties lies in the handling of the arithmetic associated with population processes. In the case studies above, we used a nominal numerical scale with successive points on it, corresponding to the ordinal sequence of states of a component.

The automatic choice may be a linear scale, i.e. the increment from a low to medium is the same as that from a medium to high, etc. A linear scale was used in the termite case study. In many instances, however, a logarithmic scale is likely to be more workable for three reasons. First, as the population becomes larger the intervals in our scale should perhaps be correspondingly larger so that concepts dealing with rates of change, such as 'increase' ($+1$), have a consistent meaning at all population sizes. Second, a linear scale with sufficient resolution at low population densities would lead to a very large number of states if the full range of population size is to be represented. Third, in describing population processes, one very often requires more resolution at one end of the scale, for example armyworm mortality (Tables 9.4 and 9.5). In addition, logarithmic scales also simplify calculations; states of the rate components (mortality, fecundity) correspond directly to movements up and down the population size scale.

In the armyworm case study a full description of the calculations is given. A logarithmic scale of base e was used because some quantitative estimation of mortality and fecundity were possible, and a scale of base e allowed most population processes to be described by about four to six states. The choice of base affects, for example, the rules for adding two populations (Table 9.8). With a base e, adding a population size of state N to a population size of state $N - 1$ or lower, results in a population size of state N. Such rounding in calculations is unavoidable and it is important that rules which govern the rounding are consistent.

The need to resort to numerical scales may appear to defeat the object of the qualitative approach. However, the mechanics of scaling and computation can remain very much in the background. The model can still be conceived and discussed purely in terms of qualitative structures.

Validation

Validation of rule-based models suffers from all the same problems as conventional simulation models. With how many data sets should the model be compared? What is a good fit? Whilst statistical techniques are sometimes used to compare 'observed' and 'predicted' in conventional simulation, this is arguably less appropriate when a model simulates systems dynamics in relative terms. A subjective visual comparison is likely to offer sufficient insight into the behaviour of the model for most purposes. One point to remember in this respect is that the assumptions about the scaling of population size alter one's perceptions considerably when making such comparisons.

More complex conventional simulation models can very often yield the same result from a number of different parameter value combinations. A systematic analysis of the reasons why there are discrepancies between model predictions and observed data can be a lengthy or impossible task. The simplification imposed by a rule-based approach may reduce the number of possible solutions or at least make the task of identifying potentially wrong assumptions easier.

Sensitivity Analysis

In conventional models the response of the model to changes in parameter values gives an indication of which parameters have most impact on model behaviour. Sensitivity analysis can be applied to a particular conclusion from the model (for example, a management recommendation). The extent to which model parameters can be changed without altering the particular conclusion gives a measure of the robustness of the conclusion (Norton *et al.*, 1983).

With a rule-based model, parameters do not exist in the same sense as in conventional models. Instead, relationships between model components are altered by modifying the appropriate set of rules to reflect alternative assumptions. For example, let us return to the set of five rules given at the beginning of this chapter which specify the relationship between plant density and seed production. To explore the question 'what if intra-specific competition has no impact on seed production?', we might change the fifth rule to read: If plant density is very high then seed production per plant is very high. Sensitivity analysis, therefore, can be closely linked to the actual questions managers wish to ask. All members of inter-disciplinary teams can participate not only in model construction, but also in formulating 'what-if' questions to evaluate model conclusions.

REFERENCES

Holt, J., Cheng, J.A. and Norton, G.A. (1990) A systems analysis approach to brown planthopper control on rice in Zhejiang Province, China. III. An expert system for making recommendations. *Journal of Applied Ecology* 27, 113–122.

Jones, T.H., Young, J.E.B., Norton, G.A. and Mumford, J.D. (1990) An expert system for the management of wheat bulb fly, *Delia coarctata*, (Diptera: Anthomyiidae) in the United Kingdom. *Journal of Economic Entomology* 83, 2065–2072.

Knight, J.D. and Mumford, J.D. (1992) The use of rule based models in crop protection. *Proceedings 1992 Brighton Crop Protection Conference*, pp. 981–988.

Norton, G.A., Sutherst, R.W. and Maywald, G.F. (1983) A framework for integrating control methods against the cattle tick, *Boophilus microplus*, in Australia. *Journal of Applied Ecology* 20, 489–505.

Starfield, A.M. and Bleloch, A.L. (1986) *Building Models for Conservation and Wildlife Management.* Macmillan, New York, 253 pp.

10 Modelling with Spreadsheets

J.D. MUMFORD[1] AND J. HOLT[2]

[1]Department of Biology, Imperial College at Silwood Park, Ascot SL5 7PY, UK:
[2]Natural Resources Institute, Central Avenue, Chatham Maritime, Chatham ME4 4TB, UK

INTRODUCTION

The use of spreadsheets for financial modelling is common in business applications. Recently, some examples of biological models developed on spreadsheets have begun to appear (Costello *et al*, 1991; Waller, 1992). Spreadsheets are also used very widely for the storage and management of biological data. Spreadsheets offer an easy way to build computer models. Even for those new to spreadsheets, useful progress can be achieved in a fraction of the time required to learn a programming language.

What is a Spreadsheet?

A large range of proprietary spreadsheet software is available (e.g. Lotus, QuattroPro, Supercalc, Excell). All spreadsheet software is relatively inexpensive and it has become an essential computing tool for decision makers. In this chapter, the Lotus syntax is used, as an example of a product available to almost everyone with access to a basic IBM compatible desktop computer.

The underlying principles of spreadsheets are common to all products. They consist of a tabular arrangement of cells, each cell having a particular address designated by its column position (usually a letter: A, B, C, . . . Z, AA, AB, etc.) and its row position (usually a number: 1, 2, etc.). A cell can store text, numerical values, or formulae. Formulae define numerical or logical operations on values or formulae in other cells. Some of the features of spreadsheets are illustrated by a simple model describing the logistic growth of a plant disease infection:

$$N_t = N_{t-1}[1 + r(1 - N_{t-1}/K)]$$

where N_t is the percentage of plants infected at time t, r is the rate of spread of the disease and K is the saturation level (all plants infected). A model such as this could be set out on a spreadsheet in the following way. The names of the parameters could be placed in column

	A	B	C	D
1			t	N_t
2	r	1.1	0	(B4)
3	K	100	(C2+1)	D2*(1+B$2*(1-D2/B$3))
4	N_0	0.0	(C3+1)	D3*(1+B$2*(1-D3/B$3))
5			(C4+1)	D4*(1+B$2*(1-D4/B$3))
6			etc.	etc.

Fig. 10.1. An example of the structure of a simple spreadsheet.

A, parameter values in column B, formulae which calculate the values of t in column C, and formulae which calculate N_t in column D.

Where formulae are entered into the spreadsheet the calculated value of the formula is displayed, e.g. in cell C3, the number 1 is displayed and in cell D3 the number 2.089 [the solution to $(1 \times (1 + 1.1 \times [1 - 1/100]))$ is displayed. Formulae do not need to be entered individually into each cell. Having been entered once (in cells C3 and D3), the formulae are simply copied for as many rows as desired. Cell references contained within formulae are automatically incremented during the copying process. To fix a cell reference during copying, so that the value of r in cell B2 is used in all cases, the '$' symbol is used when naming the cell (when the cell reference B$2 is copied to a cell below, the row reference remains fixed).

Advantages and Disadvantages of Spreadsheets for Modelling

Advantages:

- Spreadsheet software is widely available, relatively inexpensive and runs on a wide range of computers.
- Many decision makers and researchers are already familiar with the use of spreadsheets through more conventional applications.
- The basics of spreadsheet modelling can be learnt very quickly.
- Model changes are implemented immediately, rather than having to explicitly recompile the program after each change.
- Graphical displays, printing and file handling are all built-in features.

Disadvantages:

- It is more difficult to see the model structure and assumptions than if a structured programming language is used.
- Complex models can take up a great deal of memory depending upon how they are represented.
- Even moderately complex models can be rather slow to run.
- It is not easy to set up automated series of runs, for example where one might vary a number of parameters over specific ranges to evaluate the sensitivity of model output to changes in the assumptions of the model (although this can be done using 'macros').

- They are less easy to compile as 'stand alone' software for distribution than if a programming language is used.

Suitable Applications

In pest management, spreadsheet models are ideal for quick, exploratory models to get a feel for the range of likely outcomes from possible control options. They are particularly useful in a workshop setting where ideas can be tested immediately on a spreadsheet model. A simple spreadsheet model can be designed and built by participants during the course of a few hours during a workshop session. The results of a simple model can then form the basis for further discussion, analysis and experimentation.

Both quantitative simulation models and simulations with rule-based models can be implemented on spreadsheets. Examples of each of these are shown below. The first model is a quantitative simulation of a cocoa pod borer population. It is described in some detail to explain most of the basic features of spreadsheets and to illustrate an approach to spreadsheet model building. The model incorporates the dynamics of the pest and the crop as well as calculating the net returns from the crop taking into account pest damage and control costs. The second example explains the use of conditional logic in spreadsheets.

EXAMPLES

Cocoa Pod Borer Simulation Model

The cocoa pod borer (*Conopomorpha cramerella*) is a lepidopterous pest of cocoa in South East Asia (Mumford and Ho, 1988). The female moth lays eggs on the surface of immature cocoa pods. Two to three days later the larvae hatch and bore into the pod where they feed for about three weeks. Crop loss occurs from several effects of larval feeding: beans become clumped together, making their extraction from the pod difficult or impossible; pods appear to ripen prematurely, so unripe pods are harvested; and, beans are sometimes of reduced weight and quality. Mature larvae emerge from the pods and pupate on leaves in the canopy and on the ground. The life cycle can be summarized as in the upper part of Fig. 10.2. In a simplified model of this pest we can consider three natural processes affecting the rate of growth of the pest population: egg survival (which is density independent); larval survival (which depends indirectly on the density of pods, since when pods are most abundant there are more pods of suitable age for larval development); and fecundity.

Cocoa produces pods throughout the year, but in a six-month cycle. This cycle affects the survival of pod borer larvae. The crop cycle is represented in the lower part of Fig. 10.2.

Several management options are available to cocoa growers: during the low crop period managers can strip pods to manipulate the pod cycle; adults can be killed by insecticide sprays (at various efficiencies); resistant cocoa varieties could reduce larval survival; fecundity could be reduced by trapping or mating disruption using pheromones; egg survival could be reduced using parasites; etc. Managers ask questions about the relative effectiveness and cost of these options and these questions can be answered by a simple model. Each of the potential controls mentioned above could be included in the model we intend to construct.

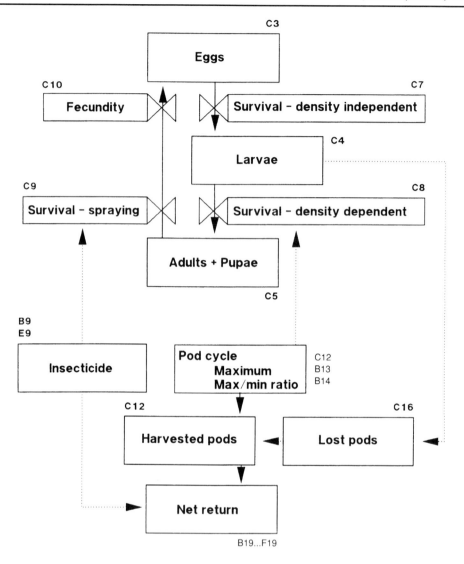

Fig. 10.2. Cocoa pod borer model (references beside each component indicate the relevant cells in the spreadsheet model presented in Figs 10.3 and 10.4).

The spreadsheet model incorporates all of the components of the system shown in Fig. 10.2 into equations in specific cells. The basic formulae for this example are shown in Figs 10.3 and 10.4. To make the model on your own spreadsheet enter the values shown in these figures into the cells indicated. Once these are entered, copy cells C1 ... C16 to D1 ... AL16 (following instructions for your spreadsheet program). You will then have a spreadsheet with 36 columns for monthly values of the crop and pod borers. Notice, in this case, that the time intervals are represented by columns rather than rows, as in Fig. 10.1. Using the graph command, you should make a graph in which the x axis is the block C1 ... AL1 (months) and two y axis series are C12 ... AL12 (harvested pods) and C16 ... AL16 (damaged/lost pods).

	A	B	C
1	Month		(B1+1)
2		Initial	
3	Eggs		(B5*B10)
4	Larvae		@INT(C3*C7)
5	Adults and pupae	10	@INT(C4*C8*C9)
6			
7	Egg survival	0.1	(B7)
8	Larval survival	0.4	($B8*C12/$B13)
9	Spray survival	1.0	@IF(C12<($B13/($B14-1)),$B9,1)
10	Fecundity	50	(B10)
11			
12	Harvested pods		@INT(($B13*(((@COS((C1-1)*@PI/3)+1)/2)*(1-(1/$B14))))+($B13/$B14))
13	Pods maximum/ha	3000	
14	Pod max/min ratio	3	
15			
16	Damaged pods/ha		@IF((C4/C12)<1,0,@INT((C12-(C12*C12/C4))))

Fig. 10.3. Contents of cells A1 to C16 in the cocoa pod borer spreadsheet simulation model.

	A	B	C	D	E	F
18	Cost and returns total	OK pods	Damaged	Gross $	Spray $	Net $
19		@SUM(C12..AL12)-C19	@SUM(C16..AL16)	(B19*0.1)	@IF(B9<1,900,0)	(D19-E19)

Fig. 10.4. Contents of cells A18 to F19 in the cocoa pod borer spreadsheet simulation model.

The cells in row 1 show time in months. The model has a monthly increment.

The cells in rows 3–5 are the monthly values of the three population stages included in the model. They are started by an initial immigration and continue based on the survival and fecundity values in rows 7–10. Egg survival is constant (0.1). Larval survival is dependent on the number of mature pods (the value in B8 is the maximum survival). As the crop nears its peak survival is high because many pods are very suitable for larval development; when the crop is at the bottom of its cycle, most pods are not suitable for larval development. The spray survival value (B9, set initially at 1.0, in which case there is no spraying and all adults survive to reproduce) is imposed only in the month when the crop cycle is at its lowest value. This is the most effective time to spray to reduce damage in the peak crop period.

The cells in rows 12–14 determine the crop production cycle. Cell B13 is the maximum number of pods harvested per hectare in a month, and cell B14 gives the amplitude of the crop cycle. The formula in cells C12 . . . AL12 produces a cycle with a six month frequency starting at a high point in month 1, fluctuating between values of B13 and B13/B14.

Row 16 gives the number of pods that are lost. This number is much less than the number that are infested, since it is assumed that no pods are lost if there is less than 1 larva per pod. Once there are more than 1 larva per pod, the loss increases exponentially until all pods are lost.

Rows 18 and 19 provide summaries of the production, losses and costs over 36 months. Cell E19 assumes that spraying costs $150 per season, if it is done.

Figure 10.5 illustrates the graphical output of the model with the original values, which assume no control.

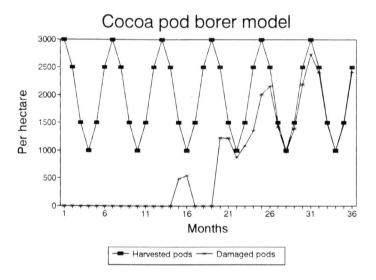

	A	B	C	D	E	F
18	Cost and returns total	OK pods	Damaged	Gross $	Spray $	Net $
19		44615	27369	4461.50	0	4461.50

Fig. 10.5. Sample output of the cocoa pod borer model (inputs are as in column B of Fig. 10.3).

The model can be used to answer management questions. Modify cell B8 to change the degree of resistance in the cocoa variety (softer pod varieties will have a high value, hard pod varieties have a low value). Spraying can be imposed with different levels of efficiency (cell B9). Pheromone trapping or mating disruption would affect the fecundity value in cell B10 (trap/confuse). Practices which affect the amplitude of the crop cycle (for instance stripping off pods in the low crop season, or synchronizing the crop with flowering hormones) can be modelled by changing the value for B14.

This relatively simple model highlights what may seem to be some counterintuitive results. For instance, removing pods in the low crop period (by increasing the value of cell

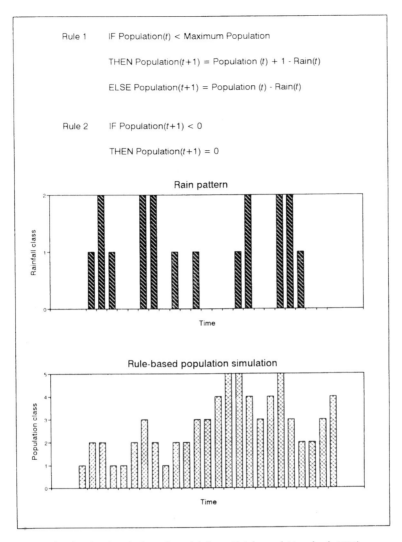

Fig. 10.6. An example of a simple rule-based model (from Knight and Mumford, 1992).

B14 to 4, equivalent to stripping out 250 pods/ha in each low crop period) pays for itself by increasing density-dependent mortality of larvae. The model also allows a manager to see the value of several partial controls applied together (for instance, some small amount of host crop resistance, B8 = 0.39, combined with some relatively inefficient spraying, B9 = 0.8). Other combinations can be tested and used to suggest either management guidelines, or further questions for field experimentation.

Rule-based Model

Knight and Mumford (1992) present a very simple rule-based model made using a spreadsheet. It is a hypothetical example of a pest population that increases by steps unless it has already reached a maximum value or is cut back by rainfall. The population has six size classes (0–5), and rainfall is in three classes (0–2). The population increments by one size class in each time period, and is reduced by one size class for each unit of rainfall. The population cannot fall to less than zero. Figure 10.6 shows an example of the output of the model with one particular rainfall pattern.

	A	B	C
1	Rule-based model example		
2	Time		(B3+1)
3	Population class		@IF(B5<5,(@IF((B5+1-B7)>=-1,(B5+1-B7),1)),(B5-B7))
4	Rain class		0

Fig. 10.7. The initial cells for the rule-based model (rain class values in row 4 must be entered for each time interval, but formulae in rows 2 and 3 can be copied to columns D to AA).

The model is set up very simply on a spreadsheet as shown in Fig. 10.7. Cells C3 ... C7 are copied to cells D3 ... AA7 to give a run of 25 time periods. The rules are specified by the @IF(condition, true value, false value) statement in row 5.

CONCLUSION

Spreadsheets provide an inexpensive tool for making preliminary models as a basis for further investigation. They give good graphical and numerical output with only a small investment of time. Both simulations and rule-based models can be made using spreadsheets.

A very limited introduction to the capabilities of spreadsheets in pest management modelling has been given in this chapter. More advanced features include a wide range of statistical, mathematical and logical functions which are of use for model building. In addition, a macro facility built into most spreadsheets allows sequences of commands to be stored, analogous to programming languages such as BASIC, PASCAL, etc.

REFERENCES

Costello, T.A., Costello, J.V., Van Devender, K.W. and Fergusson, J.A. (1991) Spreadsheet-based user-interface for a crop model. *Computers and Electronics in Agriculture* 5, 315–326.

Knight, J.D. and Mumford, J.D. (1992) The use of rule based models in crop protection. *Proceedings 1992 Brighton Crop Protection Conference*, pp. 981–988.

Mumford, J.D. and Ho, S.H. (1988) Control of the cocoa pod borer (*Conopomorpha cramerella*). *Cocoa Growers' Bulletin* 40, 19–29.

Waller, M.E.J. (1992) Spreadsheets as research tools and decision aids for cereal aphid control. *Proceedings 1992 Brighton Crop Protection Conference*, pp. 581–586.

11 Expert Systems

J.D. Mumford[1] AND G.A. Norton[2]

[1]*Department of Biology, Imperial College at Silwood Park, Ascot SL5 7PY, UK:*
[2]*Cooperative Research Centre for Tropical Pest Management, University of Queensland, Brisbane, QLD 4072, Australia*

INTRODUCTION

Expert systems are programs that mimic the processes employed by a human expert in diagnosing a problem and giving advice. Usually, expert systems are computer based, although for practical implementation in developing countries it is often more appropriate to develop the expert system on a computer and then to transfer it to a paper or manual format, for instance as a decision chart. Even without computers, the techniques of expert system design can be used to develop better decision rules. Early applications of these systems were in medical diagnosis and mining (Forsyth, 1984); more recently expert systems and related hypertext systems have been developed for use in agriculture and pest control (Peart *et al.*, 1986; Stone *et al.*, 1986; Coulson and Saunders, 1987; Bouchard *et al.*, 1989; MICROHERB, 1989; Mumford and Norton, 1989; Saunders *et al.*, 1989; Jones *et al.*, 1990; Wilkin *et al.*, 1990; Compton *et al.*, 1992; Edwards-Jones *et al.*, 1992; Knight *et al.*, 1992; Warwick *et al.*, 1992). The first knowledge-based system used in crop protection in Britain was Counsellor, which advised users about the control of cereal diseases (Jones *et al.*, 1984).

Expert systems are one group of the general set of knowledge-based systems, programs that store, process and disseminate knowledge. Decision support systems can be considered as a more general class of knowledge-based systems that provide additional, non-expert, functions, such as providing background information and performing calculations to help support a decision made by the user.

ROLE OF KNOWLEDGE-BASED SYSTEMS

Knowledge-based systems in general can perform four main roles in pest management (Mumford and Norton, 1989):

Practical problem solving – diagnosing problems and prescribing appropriate treatment.

Information processing/provision – historical databases and the presentation of future scenarios.

Problem structuring – especially the integration of disciplines and problem components, to specify key research and extension requirements for improved management.

Training – games and simulations that give experience in making decisions about specific pest problems.

To consider the role that expert systems can play in pest management problem solving, let us first reconsider the processes involved in pest management decision making, as described in earlier chapters. We saw that six processes were involved in making a decision or a recommendation:

- Pest identification
- Assessment of the level of pest attack
- Assessment of the level of crop loss
- Determination of available options and their effectiveness
- Cost/benefit assessment
- Determination of other objectives and constraints

Expert systems can be employed to help decision makers in all of these problem-solving processes, although most existing systems only address one or a few of these processes.

Many expert systems help the user to identify pests, either by keys (CABIkey, 1990), photographs in manuals (Wilkin et al., 1990), or descriptions (Compton et al., 1992). The system developed by Compton et al. (1992) helps the user to first identify 'key pests' whose control dominates that of other pests. Once a key pest has been identified, there is no practical need to identify other pests. Various methods are used to assess levels of pest attack and crop loss, either in classes (Jones et al., 1990; Knight et al., 1992) or as a numeric value (Wilkin et al., 1990). Edwards-Jones et al. (1992) and MICRO-HERB (1989) provide examples of expert systems that are mainly intended to present the decision maker with the entire list of available options for a particular problem (herbicide choice for weeds). Costs of control are calculated by many programs (Jones et al., 1990; Wilkin et al., 1990), but benefits are often not determined, either because it is simply assumed that controls should be used regardless of how well they perform (for instance, weed control (Edwards-Jones et al., 1992)), or because controls are assumed to prevent all the damage (Wilkin et al., 1990). Some programs also check to see if there are other constraints on the decision maker, such as equipment available for control (Wilkin et al., 1990).

Expert systems can be attached to database systems to help process and provide information for managers. The FLYPAST system for aphid forecasting (Knight et al., 1992) uses historical records to help decision makers give advice on the future risk of aphid attacks in England (see Chapter 13 for further details).

Building expert systems can also play an important role in helping to structure problems, forcing a degree of rigour on scientists' thinking. The prototype expert system developed by Compton et al. (1992) made the divergent views of over a dozen pest control experts explicit and provided a channel through which they could reach some common advisory rules for tropical grain storage pests.

Expert systems can also be used as training tools, by providing information and by allowing users to test themselves against the rules used in the system. The hypertext facility in some expert system programs is especially useful for training, since the depth of information provided can be determined by the user. Warwick *et al.* (1992) used hypertext to present information in a manual on pesticide-related law in the United Kingdom. Users can find summaries of legislation relevant to use, manufacture, transport, etc. of pesticides, with cross references to related legislation in additional hypertext.

Expert systems do not always involve a computer. The rules of an expert system can be presented as a decision tree on a wall chart and used at the farm level, where there is no access to computers. The development of such charts is often made easier by using a computer program to develop the expert system, and then transferring it to the chart, as was done for a brown planthopper expert system in Malaysia (Anon., 1990).

Expert systems can also take the form of a series of matrices leading to actions and priorities. An example of such an expert system is shown in Fig. 11.1. It gives guidance to commercial forest managers interpreting monitoring for attack by *Phoracantha semipunctata* on eucalyptus plantations (Cannon, 1992; Lencart, 1992). A transect of forest is

Matrix 1

	Missing trees less than or equal to 10%	Missing trees between 10 and 30%	Missing trees more than or equal to 30%
Pest attack less than or equal to 10%	0	0	0
Pest attack between 10 and 30%	1	1	2
Pest attack more than or equal to 30%	2	1	3

Matrix 2

Matrix 1 value	Age less than or equal to 6 years	Age between 6 and 12 years	Age more than 12 years
0	1	0	0
1	2	2	1
2	2	2	3
3	3	3	3

Matrix 3

Matrix 2 value	Action	Priority
0 (= minor)	Sanitary thinning	Low
1	Sanitary thinning	Medium
2	Sanitary thinning	High
3 (= severe)	Clear cut	High

Fig. 11.1. Expert system matrices for management of *Phoracantha* on eucalyptus in Portugal (from Cannon, 1992, based on Lencart, 1992). Values in the matrices reflect a combination of the likelihood of thinning giving effective control and the value of the remaining stand.

examined each autumn and the age of trees, the proportion of trees missing due to earlier sanitary thinning and the proportion showing signs of attack by *Phoracantha* are noted. Using the matrices shown, managers can decide among the several options and can assign a priority to them. In practice, low priority control actions are not undertaken. This system could be used either with or without a computer.

The general steps involved in building a knowledge-based system are described below.

STEPS IN DEVELOPING AN EXPERT SYSTEM

Building an expert system involves three steps: problem structuring, knowledge acquisition, and knowledge engineering and program encoding. Following these stages, careful testing of the program must be done to ensure that correct solutions are produced.

Problem Structuring

To specify what is required of an expert system, it is important to fully describe the problem to be addressed: the nature of the problem, the options available, and the users and their objectives. The more time spent in the early stages defining the problem area and how solutions need to be presented, the easier designing the system will be.

Before designing a system it is important to determine the needs and background of the user (person operating the system and giving advice) and the end-user (the grower acting on the advice given). A good program is tailored to specific users and end-users and their needs. Questions that needed to be answered include: 'How much knowledge about the subject will the users have?'; 'How familiar are users with computers?'; 'Under what conditions will the system be used? (in the office, over the phone, in the field)'; and 'What computer facilities are available? (machine type, processor speed, peripherals, etc.)'.

It is often useful to break the problem into simpler parts, and to develop an expert system in modules. This is easier to conceive as it is being built, and easier to debug and amend subsequently. The herbicide selection program by Edwards-Jones *et al.* (1992) has separate modules for selecting single herbicide applications, for sequences of applications and for tank mixes. Pre- and post-emergence products are kept in separate sets of files. The pesticide law program (Warwick *et al.*, 1992) includes details of new legislation and laws that have very wide applicability in its main program, while older and more obscure laws are kept in files that are only called as needed. Many systems have clear divisions for diagnosis of the problem, assessment of loss, and selection of recommendations (Jones *et al.*, 1990, 1993; Wilkin *et al.*, 1990; Knight *et al.*, 1992).

Knowledge Acquisition

The quality of the advice given by a knowledge-based system depends on the quality of rules and information provided by the human expert(s) who contribute to its development. A good expert must be fully conversant with the technology to be employed, understand the benefits of developing a knowledge-based system and be able to communicate his or her ideas and thought processes well. Many people who regularly make decisions do so intuitively on the basis of long experience, and it is essential to help them think through the problem in a systematic way to bring out that subconscious knowledge. This is generally done by presenting the experts with hypothetical problems and asking them to describe how they would determine the advice to give in that situation.

When many experts are consulted it is likely that some differences will arise in how they make decisions (Compton *et al.*, 1992). While this may provide some problems for the knowledge engineer (expert system builder), it can be very helpful when trying to build a robust system with a number of different paths to a recommendation. When there are different views on what advice to give, or how to reach a conclusion, it is important to encourage the experts to justify their decision-making process to other experts to test the validity of their thought processes. Compton achieved a system acceptable to 13 experts by displaying anonymous example rules and decision trees on a notice board where all the experts could see how their rules fitted into the overall grain storage system. Experts were invited to modify the rules in light of the inputs from other advisers.

Knowledge Engineering and Encoding

The rules and information that form a knowledge-based system can be organized within a computer program. This can be done either by directly programming in a computer language, such as PROLOG, C, PASCAL, etc., or even a spreadsheet program using conditional IF statements, or they can be entered using a 'shell' program.

Shells are programs that organize information and compile it into computer language. By using a shell developers can concentrate attention on the structure of the user's problem, rather than the computing problem, which then becomes secondary. There are dozens of shells on the market (Barron, 1989), but no particular software standard has emerged in this area. The shells vary in their structure, philosophy, output and ease of use. If a shell is to be used to develop an expert system it is essential to recognize this variation and to choose an appropriate shell for the requirements. For example, some shells are better at presenting information through hypertext than others, some are more adapted to lists of text, while others are more numerically or graphics based.

Generally, it is a good idea to use a shell to help structure the problem and to provide some preliminary output as the system is developed. For large applications, and where the program will be widely distributed, it is often better to develop the final version of the software in a programming language rather than a shell. Programmed systems run faster, take up less memory and are cheaper to distribute than systems written with shells. Care should be taken at an early stage not to let the shell dictate the structure of the problem.

Verification, Validation and Testing

This is a similar problem to that in simulation models. There are three steps in confirming that an expert system is working properly. This procedure is similar to that described in Chapter 6.

Verification – Do components of the system fit together logically and do what the programmer intended them to do? Discrete parts of the system should be checked against decision trees and/or rule matrices for completeness and correctness. Test runs on the parts of the system must be run to ensure that correct results are given.

Validation – Does the system as a whole give the correct advice for particular sets of conditions. This can be tested against case studies in which agreed advice would be given. The analogous situation with simulation models for validating expert systems is to ask an independent set of experts to give recommendations for some given sets of conditions. A major problem with validation of expert systems is that they are often based on subjective opinions of experts, and all experts do not agree (Compton *et al.*, 1992). In such cases, either a consensus must be negotiated (Mumford and van Hamburg, 1985), or a limited number of experts can be used (Denne, 1989; Edwards-Jones *et al.*, 1992).

Field testing –This is done with pilot users and should be conducted to compare expert systems with human experts or the best alternative form of management. Flower (1993) has tested an expert system for tobacco curing by establishing trials which are managed by the expert system alongside batches which are managed according to ordinary farm management practices. Any cases in which the expert system does not perform as well as the normal practice indicates cases in which the rules must be refined. However, care must be taken in interpreting comparisons between an expert system and human experts. Mumford and van Hamburg (1985) tested an expert system for cotton pests in South Africa on a small group of private, commercial and government cotton pest advisers. As the first recommendation was displayed to the group one adviser said that he would have made a different recommendation, but the others pointed out that his advice would be illegal according to the product label, and that the expert system was in fact correct! Cannon (1992) showed the economic value of the eucalyptus pest expert system shown in Fig. 11.1 by running a simulation model of crop growth and costs with control based on the expert system and with simply clear cutting when over 50% are attacked. The simulations covered a range of levels of pest attack, the expert system was more valuable at even low levels of attack, but was much better with higher pest incidence, since early partial control protects against some later loss.

A Worked Example of an Expert System

A Quick Guide to the General Form of an Expert System Shell

Many commercial shells can be used to construct expert system programs. The shell converts natural language (English, etc.) into computer language, which leaves the expert system developer to concentrate on the logic rather than the computer programming.

Rules in most shells are entered as IF–THEN–(ELSE) sequences. Generally, a shell program provides the user with screens with IF–THEN–ELSE templates on which to enter the rules. These templates help the user to organize the rules, and reduce the chance of errors in structure as the expert system is written. The shell should also provide a store of previously entered statements which can be recalled by number or menus, so that repetitive use of statements can be done without very much typing.

Rules are made up of 'qualifiers' and 'choices' (different terms may be used in various shells). A choice is a possible final conclusion of the system; e.g. 'spray' and 'do not spray' may be two final conclusions that an expert system could produce, so these are choices. Qualifiers form the conditions which must be met so that individual choices would be selected. For the two choices above, a qualifier might be 'pest population is', with two values, 'below threshold' and 'above threshold'.

A very simple program could have one rule:

IF	pest population is	above threshold	(qualifier, value)
THEN	spray		(choice)
ELSE	do not spray		(choice)

A rule can have more than one qualifier:

IF	pest population is	above threshold	(qualifier, value)
AND	wind is	calm	(qualifier, value)
THEN	spray		(choice)
ELSE	do not spray		(choice)

In this case both statements in the IF part of the rule must be true to select the choice 'spray'; if either of the qualifiers in the IF part has a value other than those shown in the rule, then the ELSE choice will be selected.

A qualifier can have more than one value within a condition:

IF   wind is   calm OR light   (qualifier, 2 values)

Choices in some expert system shells may be given a probability or confidence level each time they are used in a rule. Several systems may be used, for example scales of (−100 to +100), (0 to 10), or (0 and 1). For practical purposes in pest management, where it may be difficult to determine probabilities exactly, it is probably best to use a 0/1 system (definitely false/definitely true), or avoid assigning probabilities at all (which is equivalent to the 0/1 system).

Before starting to use a shell, the problem should be well thought out and described on paper. It is particularly important to consider all possible combinations of qualifier values. If some values are left out, the program may not work or the user may be confused.

It is a good idea to break down problems into smaller units to reduce the complexity of each rule. For example:

IF		bollworms are	more than 5 larvae/24 plants
	AND	crop stage is	boll burst
	AND	wind is	calm
	AND	temperature is	less than 30°C
	AND	rain is	not forecast
THEN		spray	

[note: an ELSE statement is not needed in every rule]

The example above will fail if any of the five conditions is not true (that is, a qualifier has a value other than those shown).

This rule is not very flexible, however, because there are so many combinations of the five qualifiers to consider. A more flexible way of approaching the same problem is to break it down into a series of rules with intermediate qualifiers:

1.   IF   pests   need spraying
     AND   weather is   suitable for spraying
     THEN spray

2.   IF   bollworms are   more than 5 larvae/24 plants
     AND   crop stage is   boll burst
     THEN pests   need spraying

3.   IF   wind is   calm
     AND   temperature is less than 30°C
     AND   rain is   not forecast
     THEN weather is   suitable for spraying

To answer the first part of rule 1, the program will first search other rules; in this case it would find that rule 2 might satisfy that condition. Therefore, the program would ask the user about bollworm numbers and crop stage to see if pests need spraying. It will then return to the second condition in rule 1, and search until it satisfies that condition using rule 3. Once all of rule 1 is either satisfied or proved false the program will continue to run until all other rules have been satisfied.

The flexibility of this approach can be seen by adding another pest, in a fourth rule:

4.   IF   cotton stainers are present
     AND   crop stage is   boll burst
     THEN pests   need spraying

Now the program will recommend spraying if there are either bollworms or stainers. If we had simply added the condition about cotton stainers to the long rule shown at the beginning of the example, it would only recommend spraying if there were both bollworms AND stainers.

Most shells provide prompts on the screen to show what options are available to you as you develop your rules. The shell should also cross check each new rule against the previous rules to warn about contradictions with existing rules.

Finally we will go on to develop a simple example.

Background to the Example Problem – Desert Locust Control Decisions

We will make a simple expert system to decide whether to spray desert locusts by air or ground, and add conditions to determine whether to use fixed wing aircraft or a helicopter if the spraying is by air. Information for this expert system is based loosely on *The Locust Handbook* (Steedman, 1988), which contains tables about the conditions under which various locust control strategies are suitable.

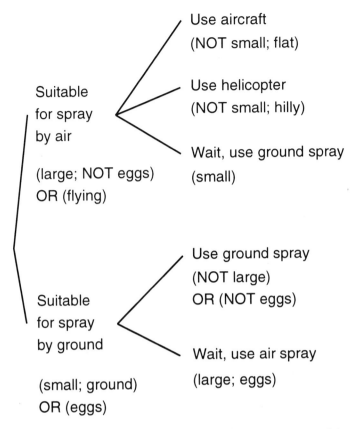

Fig. 11.2. A simple decision tree for locust control. The values of qualifiers (e.g. small, large, etc) are the same as the qualifiers and values in Table 11.2, but in abbreviated form.

The first step is to construct a decision tree (Fig. 11.2) or a rule matrix (Fig. 11.3a, b) to describe the problem. In both cases, the emphasis is on identifying pathways (sets of qualifiers and values) that lead to each of the possible choices (management actions or recommendations).

In Fig. 11.2 an initial branch in the tree is between conditions suitable for spraying by air and those suitable for spraying with ground equipment. If the locusts can be sprayed by air a further decision must be made, either to spray by fixed wing aircraft, by helicopter, or to wait until the population has landed and spray with ground equipment. For ground spraying situations, a manager must decide whether to go ahead and spray now, or to wait until the population develops and spray by air.

An alternative way to express these decision rules is to use a rule matrix, as discussed in Chapter 3. The rule matrix is easier to transfer into most expert system shells, but it does not give the sort of graphic representation of the problem that many decision makers prefer. It may be helpful to use both trees and matrices to see which is most useful in any particular case.

The rule matrix is in the form of two tables. The first (Fig. 11.3a) specifies the conditions applying to the intermediate qualifier 'suitable for spray by', which has two values 'air' and 'ground'. Note that in the rule matrix (Fig. 11.3a) some combinations

(a)

		1	2	3	4	5	6
Population	small			X		X	X
	large	X	X		X		
Behaviour	flying	X		X			
	ground		X		X	X	X
Stage	eggs				X	X	
	hopper, adult	X	X	X			X
Suitable for spray by	A=air; G=ground	A	A	A	G	G	G

(b)

		1	2	3	4	5	6	7	8	9	10	11	12	13	14
Suitable for	air	X	X	X	X	X	X								
spray by	ground							X	X	X	X	X	X	X	X
Population	small					X	X	X	X	X	X				
	large	X	X	X	X							X	X	X	X
Stage	eggs	X		X				X		X				X	X
	hop, adult		X		X	X	X		X		X	X	X		
Terrain	flat	X	X			X		X	X			X		X	
	hilly			X	X		X			X	X		X		X
Action		S A	S A	S H	S H	W G	W G	S G	S G	S G	S G	S G	S G	W A	W A

SA = Spray by aircraft; SH = by helicopter; SG = by ground; WA = Wait, spray by air;
WG = Wait, spray by ground equipment

Fig. 11.3. Rule matrix for simple locust control example: (a) conditions for intermediate qualifier, (b) conditions for final choices.

of conditions are impossible; it cannot be the case that the population is flying and in the egg stage, so those cases are not included. The results of Fig. 11.3a are then used in Fig. 11.3b, along with further qualifiers, to determine the final recommendation.

Choices

In this example there are five possible choices (management actions or recommendations).
    Spray – Aircraft
    Spray – Helicopter
    Spray – Ground
    Wait and spray by ground equipment
    Wait and spray by air

Qualifiers and their Values

There are five qualifiers used to determine the path to the various choices. These each have two or three possible values, as shown below:

Stage   eggs, hoppers, adults
Population   small, large
Behaviour   flying, ground
Terrain   flat, hilly
Suitable for spray by   air, ground

Rules

The intermediate qualifier 'Suitable for spray by' can have one of two values, determined by the rules below. It is different from the other four qualifiers, in which the values are entered directly by the user.

Rules 1-4 are derived from Fig. 11.3a to give a value to the intermediate qualifier:

1. IF   population   large
   AND   stage   NOT eggs
   THEN   suitable for spray by   air

2. IF   behaviour   flying
   THEN   suitable for spray by   air

3. IF   behaviour   NOT flying
   AND   population   NOT large
   THEN   suitable for spray by   ground

4. IF   stage   eggs
   THEN   suitable for spray by   ground

The final choices are then determined. Rules 5–10 come from Fig. 11.3b, and are illustrated in an abbreviated style.

	IF	THEN
5.	Air AND NOT small population AND flat	Spray – Aircraft
6.	Air AND NOT small population AND hilly	Spray – Helicopter
7.	Ground AND NOT large population	Spray – Ground
8.	Ground AND NOT eggs	Spray – Ground
9.	Air AND small population	Wait, spray on ground
10.	Ground AND large population AND eggs	Wait, spray by air

REFERENCES

Anon. (1990) Report of a workshop on rice pest management in the Muda area of Malaysia. Department of Agriculture, Kuala Lumpur, Malaysia, 43 pp.

Barron, R. (1989) Great expectations – Expert systems shell survey. *Systems International*, April 1989, pp. 35–40.

Bouchard, C.J., Neron, R. and Goyette, N.R. (1989) Identification of weed seedlings by parallel reasoning. *AI Applications in Natural Resource Management* 3, 71.

CABIkey (1990) Computer program produced by CAB International, Wallingford, UK.

Cannon, J.B. (1992) Cost–benefit analysis in practice: Eucalyptus and cork oak in Portugal. Unpublished MSc thesis, University of London.

Compton, J.A.F., Tyler, P.S., Mumford, J.D., Norton, G.A., Jones, T.H. and Hindmarsh, P.S. (1992) Potential for an expert system for pest control in tropical grain stores. *Tropical Science* 32, 295–303.

Coulson, R.N. and Saunders, M.C. (1987) Computer-assisted decision-making as applied to entomology. *Annual Review of Entomology* 32, 415–437.

Denne, T. (1989) An expert system for stored grain pest management. Unpublished PhD thesis, University of London.

Edwards-Jones, G., Mumford, J.D., Norton, G.A., Turner, R., Proctor, G.H. and May, M.J. (1992) A decision support system to aid weed control in sugar beet. *Computers and Electronics in Agriculture* 7, 35–46.

Flinn, P.W. and Hagstrum, D.W. (1990) An expert system for managing insect pests in stored grain. *Proceedings 5th International Working Conference on Stored-Product Protection*, Bordeaux, France, pp. 2011–2018.

Flower, K. (1993) Expert systems for agricultural research and extension: Reaping and curing flue-cured tobacco in Zimbabwe. Unpublished PhD thesis, University of London.

Forsyth, R. (1984) *Expert Systems: Principles and Case Studies*. Chapman and Hall, New York, 231 pp.

Jones, M.J., Northwood, P.J. and Crates, D. (1984) Expert computer systems and videotex as aids to cereal disease control. *Proceedings 1984 British Crop Protection Conference – Pests and Diseases*, pp. 641–646.

Jones, T.H., Young, J.E.B., Norton, G.A. and Mumford, J.D. (1990) An expert system for the management of wheat bulb fly, *Delia coarctata* (Diptera: Anthomyiidae) in the United Kingdom. *Journal of Economic Entomology* 83, 2063–2072.

Jones, T.H., Mumford, J.D., Compton, J.A.F., Norton, G.A. and Tyler, P.S. (1993) Development of an expert system for pest control in tropical grain stores. *Postharvest Biology* (in press).

Knight, J.D., Tatchell, G.M., Norton, G.A. and Harrington, R. (1992) FLYPAST. An information management system for the Rothamsted aphid database to aid pest control research and advice. *Crop Protection* 11, 419–426.

Lencart, P. (1992) Planning and follow-up system for sanitary operations against *Phoracantha semipunctata* Fab. Unpublished manuscript quoted by Cannon (1992).

MICROHERB (1989) Computer program jointly produced by Department of Agriculture for Northern Ireland, Scottish Agricultural Colleges, and Queens University, Belfast.

Mumford, J.D. and Norton G.A. (1989) Expert systems in pest management: Implementation on an international basis. *AI Applications in Natural Resource Management* 3, 67–69.

Mumford, J.D. and van Hamburg, H. (1985) *A Descriptive Analysis of Cotton Pest Management in South Africa*. Plant Protection Research Inst., Pretoria, 95 pp.

Peart, R.M., Zazueta, F.S., Jones, P., Jones, J.W. and Mishoe, J.W. (1986) Expert systems take on three tough agricultural tasks. *Agricultural Engineering*, May/June 1986, 8–10.

Saunders, M.C., Muza, A.J., Travis, J.W., Miller, B.J., Calvin, D.D., Rajotte, E.G. and Foster, M.A. (1989) Integration of pest management recommendations by an expert system. *AI Applications in Natural Resource Management* 3, 64–66.

Steedman, A. (1988) *The Locust Handbook*. Overseas Development Natural Resources Institute, Chatham, UK.

Stone, N.D., Coulson, R.N., Frisbie, R.E. and Loh, D.K. (1986) Expert systems in entomology: Three approaches to problem solving. *Bulletin of the Entomological Society of America* 32, 161–166.

Warwick, C.J., Mumford, J.D. and Norton, G.A. (1992) PESTLAW: a hypertext book on pesticide legislation in the United Kingdom. *Computers and Electronics in Agriculture* 7, 47–60.

Wilkin, D.R., Mumford, J.D. and Norton, G.A. (1990) The role of expert systems in current and future grain protection. *Proceedings 5th International Working Conference on Stored-Product Protection*, Bordeaux, France, pp. 2039–2047.

12 Pest Management Games

J.D. MUMFORD

Department of Biology, Imperial College at Silwood Park, Ascot SL5 7PY, UK

INTRODUCTION

Games can be a useful way to give decision makers quick and inexpensive experience of pest management problems. They supplement other training tools, such as lectures and practicals, are especially good for demonstrating dynamic and probabilistic events, since they allow players to see situations develop over time, and they can be repeated to illustrate different outcomes. They can also be used to teach technical information needed in decision making.

Players have a greater motivation for learning in a game format, because of the competitive element of the game. Games are fun and absorbing for players, resulting in intense participation.

However, games should only be used in small doses, as one of several alternative forms of learning. They can become tedious if too many games (or too many of the same type of game) are used during a course. It is also important to limit the length of the game to prevent it becoming too tiring; well run games can be very exhausting, both for players and trainers.

Games should be designed to give experience in a few well thought out areas, and kept short enough to focus on those subjects. On the other hand, a game must be long enough to allow sufficient replication of events to illustrate different outcomes and to see the evolution of decision making as the player gains experience.

Computers can be useful in presenting games, but they are not essential. The chance elements associated with pests can also be simulated using cards or dice, and a calculator or simple arithmetic can be used to determine the outcome of decisions.

The mechanics of playing the game must be kept very simple, so that players are not distracted from the essence of the game. Computer games should have clear on-screen instructions and menus that reduce typing inputs. Game Managers should be well rehearsed and give clear instructions so that everyone knows what they are meant to do. The objectives of the game should be explained fully at the beginning, so that players know what to expect.

EXAMPLE GAMES

We will look at three types of games: decision simulations on computer (BEET, from Imperial College, London, UK; and DIAGNOSIS, from Massey University, Palmerston North, New Zealand); biological/management simulations on computer (MI-TOR, from Ohio State University, USA; APPLESCAB, from ETH, Zurich, Switzerland; and CURACAO, from Cornell University, Ithaca, NY, USA); and a role playing game (The Green Revolution Game, by Marginal Context Ltd, Cambridge, UK).

BEET

Produced by J.D. Mumford and R.K. Day, Silwood Centre for Pest Management, Imperial College at Silwood Park, Ascot, UK (not public domain).

This game puts players in the position of English sugarbeet farmers controlling pests. The game is based on the insect and disease control problems over a 15 year period during the 1970s and 1980s in East Anglia, UK.

The purpose of the game is to:

- Experience the pest management decision-making problems farmers face.
- Learn about sugarbeet insect and disease pests and their control.
- Develop a strategy to meet farmers' objectives.
- Evaluate the limited information available to decision makers.
- Recognize the stochastic nature of pest management.
- Distinguish chance environmental effects from management influences.

The game provides:

- Opportunity to view decision makers' problems.
- Discussion of the need for and value of management information and advice.
- Opportunity to see how different strategies can be used to achieve similar results, and how strategies and success are related to individual's objectives, rather than to an absolute standard.

The game is normally played in three sessions over 3–4 hours:

Learning – The purpose of the game is explained and players are given objectives (such as maximize profit or achieve a preset average income while minimizing pest control inputs). They can ask questions about the problems they will face as decision makers. They may also search in libraries, etc. for other source material. This is active, directed learning, which is more effective than passive learning in lectures. (1 hour)

Playing – Players work in groups to ensure that they must defend each decision during the game to their partner. Several groups can play in the same room and they are encouraged to discuss their performance with their neighbours as they play, as if they were other nearby farmers. The Game Manager(s) also provide additional information and comments, and act as communication channels among the players, with roles similar to extension agents and pesticide dealers. (1–2 hours)

Summary – Each team describes their strategy and how well it worked in relation to their objectives. They discuss their perceptions of the pests, control measures and the reliability of information; the need for and value of additional information; and the technical and economic feasibility of providing further information in reality. The Game Manager(s) also discuss their view of how decision makers responded to information and how well their strategies were planned. (1 hour)

At the end of the game players have actively learned about the pest problem and its control, the decision problems faced by farmers and advisers, the value of information and the stochastic nature of pest control.

DIAGNOSIS

Produced by T.M. Stewart, Department of Plant Science, Massey University, Palmerston North, New Zealand (not public domain, available at US$60 from the author at Massey University). A windows based version of DIAGNOSIS is under development in collaboration with the Cooperative Research Centre for Tropical Pest Management, University of Queensland.

DIAGNOSIS is a decision simulation that attempts to give players practice in the art of diagnosing plant pathogens in the field. It is designed in a style similar to many computerized text-based adventure games in which players can interact with a simulated environment. In this case, players find themselves in preset plant disease scenarios where they can interview farmers, examine field conditions or plant material (by text or photographs displayed on the screen) and then make their own diagnoses on the basis of the information they have obtained. The program monitors the player's performance for later grading and provides an automatic de-briefing detailing the clues (both found and missed) which would have led the player to the correct decision.

The program is based on a generic dialogue shell which allows tutors to create many different scenarios. Text for the dialogue with the player must be written in a standard format, together with the clues specified. Digitized photographs for visual examination can be included where appropriate.

The game has four main stages:

Field – The scene is set by text on the screen. The player can examine various plant parts by using a verb–noun combination such as 'look leaves'. Most basic actions are possible. The farmer can be questioned by typing 'ask grower' and the player then specifies what he wants to ask (variety, history, fungicides, etc.). Where photographs are available these are automatically presented to the player when appropriate. Samples can be collected to take to the laboratory (leaves, soil, etc.).

Laboratory – Players are given options of laboratory tests (microscopic examination, incubation/culturing, etc.) to perform. Results of the tests are shown on request, and students can consult reference materials to help them interpret the results.

Diagnosis – When a player is ready to make a diagnosis he or she enters the name of the problem, and a few paragraphs of justification (which are saved on disk for later evaluation by a tutor).

Review – The correct diagnosis is displayed and the program lists the relevant clues that were found and those that were missed by the player along with other pertinent information.

At the end of the game the players have practised making decisions about observations, questions to ask in the field and techniques to use in the laboratory and have had immediate feedback. Further feedback is available later from tutors who have assessed the printouts of the game.

MI-TOR

Produced by J.R. Lemon and D.J. Horn, Department of Entomology, Ohio State University, Columbus, Ohio, USA (public domain).

MI-TOR is an interactive computer simulation of stochastic predator–prey processes with management inputs. The model is intentionally simplified to focus on the main issues of the game: the impact of applications of acaricide under different predator–prey scenarios.

Mites and predators are represented by symbols on the computer screen, and the populations develop on their own once play starts, apart from the application of sprays by the player.

The player can control the following factors (generally through input and utility menus at the start of a run):

1. Initial prey and predator numbers.
2. Breeding age of each species.
3. Predator starvation time.
4. Predator diapause initiation time and proportion diapausing.
5. Predator learning.
6. Pesticide resistance.
7. Dispersal.
8. Pesticide application time and kill rate on each species.
9. The duration of the model.

Players can test concepts of predator–prey interaction, and see the effect of insecticide applications with different levels of predator and prey dispersal. Output is presented both graphically and in tabular form.

There are two versions of MI-TOR. M8 is the general version with only one mite–predator system on the screen at a time. D4 is a version which includes mite and predator dispersal. Four mite systems appear on the screen at the same time; the player can select from the input menu whether mites and/or predators can disperse between the 'leaves', and whether there should be a refuge area where predators cannot feed.

APPLESCAB

Produced by P. Blaise, Institut fur Pflanzenwissenschaften, Gruppe Phytomedizin/Pathologie, ETH-Zentrum/LFW, CH-8092 Zurich, Switzerland (not in public domain, available at nominal cost from ETH).

APPLESCAB is a teaching tool which simulates the development of an apple scab epidemic in an orchard. Players can test their management abilities by trying to control epidemics with fungicide applications under a number of different conditions.

This game can be used for the following objectives, depending on the conditions selected at the beginning of the game:

1. Influence of weather on the incidence of apple scab infection.
2. Importance of coordinating pesticide application with weather and pathogen biology.
3. How spray strategies must vary according to the characteristics of the fungicide.
4. How low levels of disease resistance can be combined with fungicide sprays for apple scab control.
5. Fungicide resistance management.
6. How apple scab control in one season affects subsequent seasons.
7. Optimization of cost-effectiveness of sprays.

At the start of a run the player can control the general temperature and rainfall levels for the season, the level of primary inoculum, fungicide resistance level, apple cultivar and apple price.

The season increments daily until harvest, and one of four fungicide sprays can be applied at any time. The player is given continuous reports on weather, lesion formation, ascospores and crop stage.

At the end of the season a summary report is presented, with details of fungicides used and total costs, production and profit, and weather conditions throughout the season. Players can test different control strategies under various weather, cost and performance regimes.

CURACAO

Produced by A.J. Sawyer, Z. Feng, C.W. Hoy, R.L. James, S.E. Naranjo, S.E. Webb and C. Welty, Department of Plant Pathology, Cornell University, Ithaca, New York, USA. It has been adapted for Microsoft Windows by P.A. Arneson and B.E. Ticknor (available through B.E. Ticknor at no charge provided all use is registered and evaluations are sent to the authors).

CURACAO simulates the eradication of the screwworm from the Caribbean island of Curacao, the earliest successful use of the sterile insect release technique to eradicate a pest. Its objectives are to give students an understanding of the principles of using sterile insects for eradication and to allow them to test the assumptions of the control model.

The following factors can be altered in the simulation to test their impact on the rate at which eradication is achieved:

1. Spatial resolution and emigration to other parts of the island.
2. Order of events (i.e. whether release of sterile insects and mating both occur before dispersal).
3. Density dependence (mating, fecundity, survival, emigration).
4. Stochastic or deterministic events.
5. Competitiveness of sterile males vs. wild males.
6. Spatial heterogeneity.
7. Accuracy of the estimation of the native insect population size.
8. Release rate of sterile insects.

The program presents a graphic representation of the screwworm population on the island at the designated resolution. Players see the population change over time, with the objective being to learn how to achieve eradication within the shortest time with minimal inputs of sterile flies.

The Green Revolution Game

Produced by G.P. Chapman and E.A. Dowler, Marginal Context Ltd, 36 St Andrew's Road, Cambridge, UK.

The game is a simulation of the impact of Green Revolution technologies on a village in northeast India. The players take the role of farmers growing rice under the uncertainties of weather and changing prices. Each farmer in the village must decide whether to adopt new technologies: high yielding rice varieties, fertilizer, pesticide, and tube wells.

Drought and pests occur at random (determined by drawing cards), and prices and social structure change as a consequence of the adoption of technology, chance events and group interactions.

The game can be played by 12–24 players and about six growing seasons can be played during one full day session, with an explanatory session at the start and a summary session at the end (about 1 hour each).

This game concentrates on the social and economic impact of technology, and therefore the pest control component is quite superficial. However, the game illustrates the role of pest management as an integral part of rural development programmes, and the important role of risk avoidance in adoption of technologies like spraying.

BIBLIOGRAPHY

Blaise, P., Arneson, P.A. and Gessler, C. (1987) APPLESCAB, a teaching aid on microcomputers. *Plant Disease* 78, 574–578.

Chapman, G.P. (1973) The Green Revolution: a gaming simulation. *Area* 5, 129–140.

Chapman, G.P. (1982) *The Green Revolution Game Managers' Handbook*. Marginal Context Ltd, Cambridge, UK. 36 pp.

Lemon, J.R. and Horn, D.J. (1989) MI-TOR. Project Document 89-2, Ohio Pest Management and Survey Program, Ohio State University, Columbus, Ohio, USA. 24 pp.

Mumford, J.D. (1984) A computerised decision model of sugar beet insect and disease control. *Proceedings 1984 British Crop Protection Conference*, pp. 615–619.

Sawyer, A.J., Feng, Z., Hoy, C.W., James, R.L., Naranjo, S.E., Webb, S.E. and Welty, C. (1987) Instructional simulation: Sterile insect release method with spatial and random effects. *Bulletin of the Entomological Society of America* 33, 182–190.

Sawyer, A.J., Feng, Z., Hoy, C.W., James, R.L., Naranjo, S.E., Webb, S.E., Welty, C., Arneson, P.A. and Ticknor, B.E. (1990) A simulation of the sterile insect release method of pest control. *CURACAO User's Manual*, Cornell University, Ithaca, New York, USA. 41 pp.

Stewart, T.M. (1992) DIAGNOSIS – A microcomputer-based teaching aid. *Plant Disease* 76, 644–647.

13 Database Systems

J.D. Knight[1] and R.K. Day[2]

[1]Department of Biology, Imperial College at Silwood Park, Ascot SL5 7PY, UK:
[2]International Institute of Biological Control, PO Box 76250, Nairobi, Kenya

Introduction

This chapter introduces the use of database systems in pest management. First, two basic questions are addressed; what is a database system and what are its components? Then, two specific database systems are described to illustrate the use of databases for pest surveillance purposes. WormBase is an operational system, used in association with the forecasting and control of armyworm (*Spodoptera exempta*) in East Africa. FLYPAST is a system that has been developed in the United Kingdom for the analysis of aphid (suction trap) data for research and forecasting purposes. Finally, the procedures for planning, developing and implementing a database system are discussed.

What is a Database System?

To answer this question let us start by defining what we mean by a database: it can be defined as **a large collection of related data, organized for a particular use**. To expand on this let us now look in more detail at each of these terms:

Large – A small database can be very useful, but the advantages of computerized database technology are only realized when the volume of data becomes relatively large. Mainframe computer databases can store almost unlimited amounts of information, but technological developments in recent years have greatly increased the amount of information that can be stored in a desk-top computer. The advent of optical disks is increasing the storage capability even more dramatically allowing up to 600MB of information to be stored on a single disk.

Related – The different items of data or information in a database are related to each other, and to the particular use for which the database has been created. Although this may appear obvious, we shall see that the exact way in which different items of data are related has a special importance in database design.

Data – Although databases were originally developed to contain data in the form of numerical measurements or observations they can also be used to store and manipulate textual and qualitative data. Databases can therefore be thought of as containing information or knowledge as well as data.

Organized – Data can be organized in a database in several different ways. At one level there is the type of data model used, one of which we will be looking at in more detail. For a given data model the data can be variously organized, and this is an important part of database design.

Use – One of the main areas in database research has been to devise ways of storing and representing data that are independent of the application, so that the same data can be used for various purposes. For applications in pest management decision making, it is suggested that a database should only be developed where the potential for such a system has been clearly defined.

Database systems are in use in a wide range of applications. Some of the major types of use to which they are put are shown below, though any one system may incorporate several different uses.

File/data management systems – At the simplest level, database systems can function as electronic filing systems. They perform much the same role as paper filing systems but with improved efficiency.

Reference sources – A common use of databases is as a store for reference information, for example computerized bibliographies. This type of database is described in Chapter 15.

Historical archives – In past research programmes, large amounts of information may have been gathered. Computerization of this information can greatly improve its accessibility, and perhaps eliminate repetition of research carried out by earlier workers in a field.

Geographic information systems (GIS) – In these systems the spatial component of data sets is particularly catered for. In the following chapter, Dan Johnson covers these systems in more detail and gives examples of the application of GIS to a pest problem.

Real-time or on-line systems – Some pest management decisions, such as pest forecasting, are made on the basis of large amounts of information, including recently gathered or real-time data. The easiest way to use these data can be via a database system.

Expert database systems – Continuing software developments are resulting in convergence between traditionally distinct application areas. Thus databases are now becoming an integral part of expert systems, and several expert system shells provide facilities for accessing information directly from database files.

What are the Components of a Database System?

The starting point for any database system is the choice of a suitable software package. Although there are many different database software packages available, many of them use a similar set of commands and functions to those found in dBASE IV (Borland International). These are known as the xbase family of packages and are probably the most widely used group designed for IBM PC machines. Examples include dBASE IV 1.1 (Borland International), Clipper 5.0 (Computer Associates), FoxPro 2.0 (Microsoft), Paradox (Borland International), Quicksilver (Wordtech Systems Inc.), R:Base (Microrim (UK) Ltd). Where the term xbase is used in this chapter it refers to the family of packages, dBASE refers to the Borland International product.

Database systems contain two parts: the data and the procedures for maintaining and using the data files.

The Data

The way in which the data are stored in a database is determined by the data model used. There are a number of data models in use but we shall deal only with the relational model. Many of the proprietary database languages use the relational model.

In the relational model, all data are stored in **relations**, which appear as data tables. Each table is stored as a file in the computer. An example of a data relation, called TRAP is:

TRAP (Trap-Number, Trap-Name, Trap-Location)

This relation can be drawn as a table:

Trap-Number	Trap-Name	Trap-Location
001	Muguga	0111S 3637E
002	N.A.L.	0115S 3645E
003	Kibos	0003S 3451E

In this table the column headings are called **Fields**. The number of fields is called the degree of the relation. Each row is called a **Record**. If a field has values which uniquely identify the different records, the field is called a key field or the key of the relation. Trap-Number is a key field in the above example.

To create a database file using the relational model we therefore need to specify a relation and the field names or attributes of the relation. In xbase, the relation name is the name of the file containing the data records.

As well as naming the fields, we also have to specify the type of data that each field will contain. There are two main types of field.

Numeric – These fields contain numbers, which can be integers or real numbers.

Character – These fields contain any character string, including digits, punctuation and symbols.

In xbase, as well as the two fields above, there are three additional field types.

Date – These fields contain a date in the form 01/02/89. The date is actually stored in the computer as a Julian date (the number of days since an arbitrary starting point), but the user only sees the data as a calender date. This is a very useful field type in many applications.

Logical – A logical field contains either True or False. For example a database might contain a field indicating whether a certain species of insect was present in a particular sample. True would indicate presence of the species whilst False would indicate absence.

Memo — A memo field is an extended character field, which allows for large amounts of text to be entered.

Finally, having defined the field types, the field widths also have to be specified in xbase. This sets the maximum amount of space that an entry in a particular field can occupy. For numeric fields, the overall width of the field has to be specified, and the number of decimal places. If decimals are required, one space for the decimal point has to be included. Character fields can be any length up to a maximum of 254, date fields automatically take 8 spaces and logical fields 1 space. Memo fields are treated slightly differently and come in blocks. One space in a database field occupies one byte of storage space on the disk, so the approximate size of a data file is the product of the sum of the field widths and the number of records.

The design of a database file is thus a fairly straightforward process akin to designing a results table. However, for improved efficiency and performance of a database, the process of data **normalization** is usually desirable. There are several steps in this process, but the following example illustrates the type of procedure involved.

Suppose we have a collection of information about trap catches of a particular species of pest in a network of traps. We might construct a data table as follows.

	Trap-Number	Trap-Name	Trap-Type	Date	Number caught
Record 1	001	Muguga	Light	07/01/86	3
Record 2	001	Muguga	Light	08/01/86	2
.					
.					
.					
etc.					

This table may contain all the information that is required, but it has certain drawbacks.

1. Where a trap is counted on a number of occasions, for each record the trap name and trap type has to be repeated.
2. If the name of a trap changes, the name has to be changed in every record for that trap.
3. If there happen to be no records for a particular trap, we have no way of recording that the trap exists using the current table.

These problems can be overcome by creating separate relations for repeating groups. Thus the above table would be split into two tables as follows:

1. Trap list

Trap-Number	Trap-Name	Trap-type
001	Muguga	Light
002	N.A.L.	Light
.		
.		
.		
etc.		

2. Trap catches

Trap-Number	Date	Number caught
001	07/01/86	3
001	08/01/86	2
.		
.		
etc		

This arrangement gets rid of the three problems above, but causes another potential problem. What happens if we want to know the name of the trap stations or the location of the trap that is operating when we are looking at the trap data? This problem can be overcome by linking the two files on the common field name (Trap-Number), which in xbase is called setting a relation. In this case the file for trap catches is called the **parent** and the trap list file is called the **child**. In more complicated situations, the child could then be a parent in another relation. Once a relation has been set, the software automatically keeps track of the related information in the child files.

Normalization of the data and linking data files in this way is not compulsory, and there are occasionally situations where the general rules should be broken. However, normalization usually results in a more efficient database that is easier to maintain.

Procedures for Maintaining and Using the Data

The data or information in a database only becomes valuable when it can be easily used. There are several types of operation that are commonly found in database systems, which are described in turn below. In some database systems these procedures may be written by the programmer; in other cases, the facilities are built into the database package itself.

Data entry

Entry of data into a database is an important component of a database system. Where possible, data entry procedures should be designed to match the equivalent paper files or forms. This improves acceptance of a system by potential users and reduces the chance of errors.

Reducing the occurrence of errors in a database is an important consideration. At data entry the database system can be designed in such a way that it automatically checks for various types of errors: values out of bounds (or unusual values), duplicate data, or illogical data combinations. This checking is aimed at maintaining the **integrity** of the database.

One way of reducing errors is to provide guided data entry procedures. Thus, rather than typing in the name of a trap for example, the data item for a particular entry is selected from a menu. As well as reducing the number of keystrokes required by the operator, the content of the field is limited to that determined by the system designer. This approach is being used in the armyworm database system, described later.

A further way to speed data entry and reduce errors is by automating data entry. Various instruments for measuring environmental variables can log data electronically, which can then be loaded directly to a database. Data collected by hand can be entered into a database using a light pen to point at one of a range of bar codes on a prepared sheet; this method is used to record the aphid data in FLYPAST, which is also described later.

Data editing

Once data are in the database, even if they have been thoroughly checked at entry, occasions arise when editing is required. The new values must again be subjected to error checks. Sometimes it is necessary to prevent editing of certain critical data fields, by all users or by a particular class of users.

Data viewing/browsing

Although the main value of a database system may be in its summarizing and analysis functions, users like to be able to see the raw data. This can be combined with editing and other facilities.

Data presentation

Once data are entered in a database, presenting the data in a variety of ways is easily accomplished. Various type of presentation can be used depending on the requirements of the user. Data reporting and summarizing can be automated; for example, a monthly report can be programmed so that the appropriate printed tables can be generated simply by selecting 'Print monthly report' from a menu. Packages like dBASE provide various facilities for generating report and summary procedures. Mapping is another useful way of presenting data, which is illustrated in the systems described later.

A further database technique for providing the type of output required is **indexing**. Consider the following database fragment.

Trap-Number	Date	Number of moths caught
001	01/06/90	1
002	01/06/90	0
003	01/06/90	15
004	01/06/90	21
.		
001	02/06/90	3
002	02/06/90	42
003	02/06/90	22
004	02/06/90	1
.		
001	03/06/90	27
002	03/06/90	12
003	03/06/90	0
004	03/06/90	4
.		
etc.		

Displaying the data in this order is appropriate if we want to see what is happening at all the different traps operating on a particular date, but not if we want to look at the sequence of catches at a particular trap. One way round this is to sort the data file into a new order, but sorting is time consuming and, every time new data are added, the file would have to be resorted.

This type of problem is overcome by using index files. An index is an internal database device for recording the order in which records should appear according to criteria specified by the user. The *actual* order of the data records does not change, so when new data are added, instead of re-sorting the entire database, only the index position of the new record has to be determined.

A single file can have several indices associated with it. In the example above, if we now wish to see all the data for a single trap together, we would need to index the file on the trap number. The above data would then appear as in the following table. Indices are a very powerful tool in database systems.

Trap-Number	Date	Number of moths caught
001	01/06/90	1
001	02/06/90	3
001	03/06/90	27
.		
002	01/06/90	0
002	02/06/90	42
002	03/06/90	12
.		
003	01/06/90	15
003	02/06/90	22
003	03/06/90	0
.		
etc.		

Database queries

A major benefit of computerized database systems is that queries can be executed easily. For example, in the above file we might want to find all the trap catches that were above a certain level. The exact syntax of the command varies according to the database system being used, but the command would translate:

> SELECT all the records WHERE:
> the value in field NUMBER > 25.

Compound conditions are also possible:

> SELECT all the records WHERE:
> the value in field NUMBER > 25
> AND the value in field DATE = 01/01/89

Different database systems use different constructions to make such queries. A common query language is called SQL (structured query language) which is similar to the examples given above.

Another type of query method that can be used is Query By Example (QBE). In this case, instead of having to specify the field names and their values, the user is presented with a form on the screen of all the different data items in the database, and can then fill in the conditions in the appropriate places. In the above example under NUMBER the user would enter > 25, and under DATE the user would enter 01/01/89. The system then constructs the full query automatically.

One particular type of query that is frequently used by pest forecasters involves searching a historical database for analogous situations to the current one. The value of forecasting by analogue may be open to discussion, but a database system is well suited to such an approach.

Data analysis

In many cases a database system will be required to perform analysis of some kind on the data it contains. This analysis may be relatively simple, such as calculating the mean number of insects caught per night at a particular trap over a period of time. Combined with query conditions, quite complex analysis can be performed, using one line commands.

Data synthesis

The final stage in the utilization of data can be described as synthesis, in which the results obtained from analysis of different sets of data are brought together and used to draw conclusions or make decisions. These operations can be included in a database system, although here the distinction between a database system and some of the other labels used for computer programs becomes less distinct. The data in a database may be used to drive, test or validate models of the various kinds discussed elsewhere in this book. There are two approaches to the use of the data. First, the model can be written in a procedural language and access xbase files as required. Second, the model can be written in the xbase language itself allowing easy data manipulation and access. The interpretation and synthesis of information in a database may also require an expert system type approach if fully quantitative methods are not appropriate. Expert systems are discussed in Chapter 11.

Having presented the basic concepts, and highlighted the tasks that a database system can perform, let us now look at two specific examples to show how database systems can be usefully applied to pest management problems.

EXAMPLE PEST MANAGEMENT DATABASES

WormBase: A Database Management System for African armyworm Forecasters

Background

The African armyworm (*Spodoptera exempta*) is a noctuid causing damage to graminaceous crops and pasture in many parts of Africa, and is particularly serious in East Africa. The adult moths can migrate hundreds of kilometres on prevailing winds, during which time they would tend to be dispersed. Since airborne moths can become aggregated in convergent wind flows, such as at the edge of storms, this can lead to simultaneous oviposition, resulting in the sudden appearance of larvae at very high densities.

In 1969, a forecasting and information service was established at Nairobi, Kenya, to provide weekly situation reports and forecasts to countries in the East

African Community, enabling farmers and agricultural officers to prepare for imminent outbreaks. The service has continued since then, with national services starting in Kenya and Tanzania: the regional service has been taken over by the Desert Locust Control Organization for Eastern Africa (DLCO-EA). Forecasts are based on data from a network of light and pheromone traps and on reports of recent outbreaks, knowledge of trends in the seasonal distribution of outbreaks, and weather maps from the meteorological office.

Because the forecasters use large amounts of historical and real-time data, the potential advantages of using a database system were recognized some years ago, and an initial attempt was made to set up a database system in Nairobi. However, this was unsuccessful as the system was not directly accessible to the forecasters, and could only be operated by computer staff. With recent developments in both hardware and software, the opportunity arose to develop a system without those constraints. With funding from the Overseas Development Organisation, Natural Resources Institute (NRI), and in collaboration with DLCO-EA, NRI and the Kenyan and Tanzanian armyworm forecasting units, The Silwood Centre for Pest Management has developed WormBase, which is now in operational use in Kenya and Tanzania.

Structure of the System

WormBase is divided into three sections, accessible via a menu:

1. Database.
2. Forecasting tools.
3. Utilities.

Database

The database menu (Table 13.1) allows a range of operations to be conducted on the original data, including maintenance and utilization of the data. Where appropriate, season, country and type of data to be used are specified by menu selections, and the required operation can then be executed.

Figure 13.1 shows the output from selecting the option to plot the Kenyan light trap data for 1985/86 season and Fig. 13.2 shows the output from selecting the option to print the outbreak data for all trap areas in Kenya for February 1988. This printout has been configured to display the estimated time of the peak of egg hatch.

Forecasting tools

Although the database menu allows the user to carry out certain operations that make use of the data, there are a number of more specific uses that are provided through the forecasting tools menu. These facilities are grouped under three subheadings according to the type of data they primarily relate to.

Trap data tools – All daily trap data entered into the system are automatically summarized into a weekly record comprising the geometric mean catch and the maximum single

Table 13.1. The database menu.

Menu item	Sub-menu item	Explanation
Season		The season runs from 1 October to the following 30 September.
Country		The system is currently set up for Kenya and Tanzania. National installations can only access national data, but regional installations can access data from either country.
Data type	Light trap	Daily catches of male and female moths caught at each trap.
	Pheromone trap	Daily catches of male and female moths caught at each trap.
	Rainfall	Rainfall data from trap stations. (Meteorological office data are not recorded).
	Outbreak reports	Any reports of outbreaks, in whatever level of detail is available.
	Forecasts	Weekly forecasts issued by the forecaster (allowing automatic verification).
	Trap stations	Details of trap stations.
	Gazetteer	Details of outbreak sites, or any other locations.
Operation	Enter new data	Data entry forms match paper forms as far as possible, and menus are used extensively to limit the amount of typed data entry. Extensive error checking is included to maintain the integrity of the database.
	View/Edit	The raw data can be viewed easily, and records edited and deleted if necessary. Indexes allow the data records to be viewed in different orders as required, and searches for specific data can be made.
	Print	Selected parts of the database can be printed, with different printing formats available to suit the requirements.
	Plot	Selected parts of the database can be plotted on a map of the appropriate country, with the option of printing on a dot matrix printer.

night's catch for each trap. This file can be viewed in several index orders, allowing, for example, data for a trap to be compared with the historical data for the trap for the same period in the season. Forecasters make this comparison as a way of assessing the relative size of a trap catch, as different traps have different sensitivities. The summary data can also be plotted, allowing comparisons between seasons, and between trap types (Fig. 13.3). Other programs in this section provide tabulated summaries of trap data, and provide reports on trap reporting efficiency.

Outbreak data tools – All outbreak reports entered into seasonal files are also automatically entered into a summary file. At data entry or editing, the system calculates and records with the data the predicted egg hatching and pupal emergence dates, based on details of the report. The outbreak tools menu includes facilities to produce summary

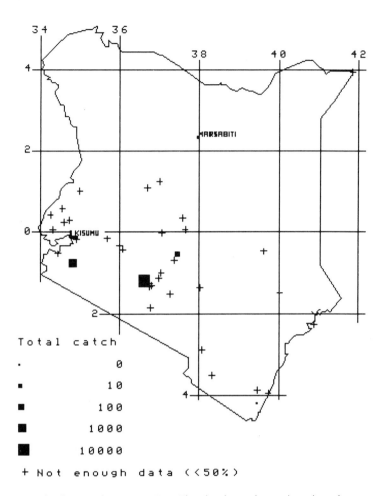

Fig. 13.1. An example of output from WormBase. The plot shows the total catches of armyworm moths in Kenyan light traps for the 1985/86 season.

maps for a single season, or over a number of seasons, showing the historical probabilities of outbreaks occurring in each half-degree square in the country. Alternatively the analysis can be conducted for particular administrative units, to provide information useful to agricultural officers planning their seasonal budgets and expenditures.

Forecasts — WormBase does not produce forecasts as such. However, as well as the facilities described under the previous sections, there are several programs that specifically assist forecast production. One provides output that can be used in manual forecasting or to run an expert system developed by the regional forecaster (Odiyo, 1990). The output includes summaries of current trap catches and outbreak reports, calculation of the probability of outbreaks occurring in specific districts for the time of year. Another program in this section allows automatic verification of forecasts by comparing the forecasts issued with the reports of outbreaks. An example of the output from this last option is shown in Fig. 13.4.

Print Specifications
Country   :   Kenya
Season   :   1987/88
Dates   :   01/02/88 to 29/02/88
Date range   :   Not included
Outbreak stage :   Egg hatching (predicted peak)
Area   :   All areas
Printing order :   District
Printed   :   04/08/92 at 16:14

Rec No.	Date Observed	District	Place	Coordinates	Stages seen	Area	Host Plant	Spray	Predicted hatching	Predicted emergence
025	=29/02/88	Bungoma			LU	11-100 ha	Crops		17/02/88 ± 5	15/03/88 ± 5
026	=29/02/88	Bungoma			LU	ha	crops		17/02/88 ± 5	15/03/88 ± 5
017	23/02/88	Busia	Kocholia	0038N 3420E	LU	11-100 ha	Both	Yes	08/02/88 ± 7	12/03/88 ± 7
018	23/02/88	Busia	Osurete	0037N 3412E	LU	ha			11/02/88 ± 5	09/03/88 ± 5
019	23/02/88	Busia	Kamolo	0000N 3400E	LU	No data			12/02/88 ± 5	09/03/88 ± 5
020	24/02/88	Busia	Khayo Matibo	0030N 3415E	LU	No data			12/02/88 ± 5	10/03/88 ± 5
021	=24/02/88	Busia			L1 L2	No data			12/02/88 ± 5	10/03/88 ± 5
016	18/02/88	Busia	Kamiriai	0040N 3418E	LU	No data	Crops		16/02/88 ± 3	14/03/88 ± 3
022	26/02/88	Kakamega	Gisambai	0001N 3445E	LU	25.0 ha	Both	Yes	14/02/88 ± 5	12/03/88 ± 5
027	05/03/88	Kitui	Ikutha	0204S 3811E	LU	No data	Crops		22/02/88 ± 5	20/03/88 ± 5
028	05/03/88	Kitui	Kanziko	0200S 3820E	LU	0.5 ha			22/02/88 ± 5	20/03/88 ± 5
029	05/03/88	Kitui	Moso	0137S 3803E	LU	No data			22/02/88 ± 5	20/03/88 ± 5
024	03/03/88	Narok	Tandui estate	0105S 3503E	L6 PU	No data			13/02/88 ± 8	11/03/88 ± 8
023	29/02/88	Narok	Tandui estate	0105S 3503E	LU	No data			17/02/88 ± 5	15/03/88 ± 5

Fig. 13.2. A printout from WormBase showing the details of outbreak reports for February 1988 and the predicted times of peak egg hatch and emergence with the possible ranges. The stages seen are: LU larva of unknown age, presumed fourth to sixth instar; L1, L2, L6, first, second and sixth instar larvae; PU pupae.

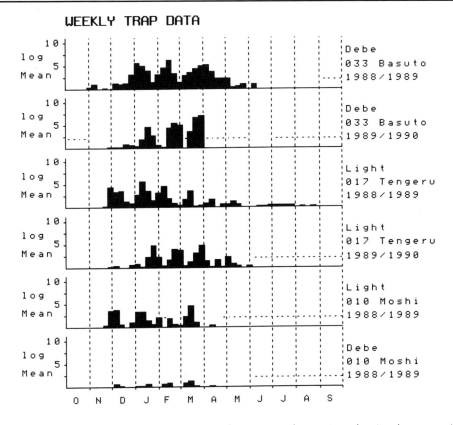

WEEKLY TRAP DATA

Fig. 13.3. Histograms from the 'plot trap summary data' option of WormBase showing the geometric mean for the weekly totals for selected traps and seasons.

Utilities

As WormBase has been written for use by staff with little or no previous computer experience, several routines have been included providing easy access to DOS operations. The principal ones are data backup and restore, which can run via menus for all the data or specific data types, as necessary.

FLYPAST: A Database Management System for Aphids in the United Kingdom

FLYPAST has been developed as a joint project between the Silwood Centre for Pest Management and the Institute of Arable Crops Research (Rothamsted, UK), to improve both the ease and speed of access to aphid suction trap data (Knight *et al.*, 1992).

Background

In Europe, aphids are one of the most important group of insect pests: some species cause direct feeding damage, whilst others are important vectors of virus diseases. A suction trap network, run by the Rothamsted Insect Survey (RIS), was established in 1964 to provide

```
FORECAST VERIFICATION FOR ALL DISTRICTS
Country             :   Kenya
Season              :   1980/1981
Forecasts           :   3 to 6
Outbreak data range :   Included
Outbreak stage      :   Egg hatching (predicted peak)
Printed             :   05/08/92 at 09:08
```

No.	Period covered	Forecast : NONE Districts	Outbreaks	Forecast : LOW Districts	Outbreaks	Forecast : MEDIUM Districts	Outbreaks	Forecast : HIGH Districts	Outbreaks	Forecast : IMPOSSIBLE Districts	Outbreaks
3	17/11/80-23/11/80	Kirinyaga * Kitui * Machakos * Murang'a * Taita-Taveta * All others									
	Totals	42	5(0)	0	0	0	0	0	0	0	0(0)
4	24/11/80-30/11/80	Kirinyaga * Kitui * Machakos * Nairobi * Murang'a * Taita-Taveta * All others									
	Totals	42	6(0)	0	0	0	0	0	0	0	0(0)
5	01/12/80-07/12/80	Kiambu * Nairobi * All others				Embu Isiolo Kirinyaga * Machakos * Meru * Murang'a * Nyeri					
	Totals	35	2(0)	0	0	7	3	0	0	0	0(0)
3	08/12/80-14/12/80	Kiambu *(P) Nairobi * All others						Embu Kajiado Kirinyaga * Kitui Machakos * Meru Murang'a * Nakuru Narok * Taita-Taveta			
	Totals	31	2(1)	0	0	0	0	11	5	0	0(0)
	Grand totals	150	15(1)	0	0	7	3	11	5	0	0(0)

(P) indicates outbreak covered by forecast for Province

Fig. 13.4. A printed table from WormBase showing the number of incorrect and correct forecasts for all districts in Kenya.

continuous data on the distribution and aerial abundance of aphids throughout the country. The data have been of considerable value in the development and provision of pest forecasts and warnings to the agricultural industry (Taylor 1973, 1977; Harrington *et al.*, 1991; Tatchell, 1991), ranging from examination of the long-term dynamics of a pest (Cammell *et al.*, 1989) to the provision of current, or real-time information to the industry to assist with pest control decisions (Bardner *et al.*, 1981; Woiwod *et al.*, 1984; Tatchell, 1985).

At present (1991) there are ten suction traps operating in England and four in Scotland. Samples are collected daily from the suction traps, identified and counted, and the data entered into a computer using a light pen to read bar codes for trap site, catch date, and aphid species (Woiwod *et al.*, 1984).

Initially, in 1971, a database was set up on the Rothamsted mainframe computer, to store the large amount of information being recorded. It was updated once a year from record books of catches. In the 1980s, a database system was set up on a microcomputer, to allow summaries of the aphid catches to be produced (Woiwod *et al.*, 1984).

Development of FLYPAST

Initially, a system specification was drawn up, based on the type of queries that were currently made and could potentially be made of the data. This was used as the basis for developing menu-driven programs, to provide predefined tabular and graphical summaries, and to run models within decision support systems. The structure of the system and the principal menus are shown in Fig. 13.5.

FLYPAST is designed to run on an IBM compatible PC with 640K memory, MS-DOS 3.3 (or higher), VGA screen and a hard disk drive. The database was developed

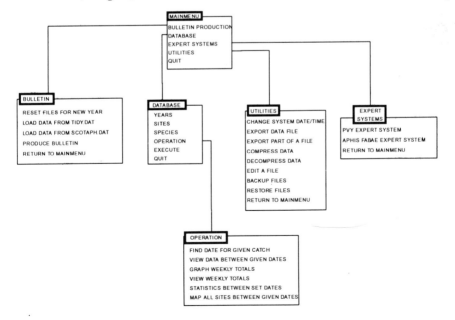

Fig. 13.5. The structure of FLYPAST and the principal menus. By selecting the appropriate option from the menus the user can progress to the part of the program that they wish to use.

Table 13.2. Descriptions of the predefined information output from FLYPAST and the way in which it can be used for research and forecasting.

View the daily catches between selected dates

Output: Daily catch information, such as the number of aphids recorded, the number of days over which the sample was collected, and the status of the trap (Fig. 13.6)

Use: Simple comparisons of the total number of a species and either females or males, can be made between traps and specified dates. The numbers of aphids observed on crops can be related to the numbers recorded in nearby suction traps for interpretation of trap catches for a particular area

Graph or view the weekly totals of aphids caught

Output: The weekly totals of aphids and of males alone for the specified trap(s), year(s) and species are presented as a table or a bar chart

Use: This information may be used to compare the size and timing of migrations of species at different sites and/or years, and is particularly useful in the preliminary exploration of data to identify changing migration patterns (Fig. 13.7)

Statistics of aphids caught between selected dates

Output: Total catch size, mean number of aphids trapped per day or the mean number per catch and the maximum number of aphids sampled during the specified period (Fig. 13.8)

Use: Allows rapid recall of numbers of aphids caught or the day on which most aphids were caught in a given period. This information can be used to explore the relationships between weather and migration

Map the total catches at each site between selected dates

Output: Total numbers of aphids recorded are plotted as bars on a map of the British Isles at the approximate location of the trap (Fig. 13.9), so giving an impression of the spatial distribution of the species

Use: Data for consecutive weeks, months or years, can be mapped to visualize how populations are developing across the country, or how quickly a new species spreads, and any changes identified rapidly

Retrieval of the date of the nth catch

Output: The date on which the first, or any other number of aphids, was caught can be extracted from the database

Use: Data for specific phenological events, particularly the date on which a species is first recorded in a year, are of particular value for forecasting

using dBASE IV version 1.1 (Borland International) with graphics functions provided by dGE (Bits per Second Ltd), but a fully compiled version using FoxPro 2.0 (Microsoft) has subsequently been produced.

Function of FLYPAST

To meet the needs of various users FLYPAST can process both historical and real-time (suction trap) information within three contexts:

FLYPAST
Accessed on 22/08/91 at 13:37:44

Site: WYE KENT Year: 1989 Catches recorded between 1 JANUARY AND 31 JULY						
Start date	End date	Number Caught Males & Females	Males Caught	Cumulative Males & Females	Cumulative Males	Trap Status
Species: P. humuli						
19/5	19/5	1		1		Normal
20/5	21/5	3		4		Normal
22/5	22/5	8		12		Normal
23/5	23/5	14		26		Normal
24/5	24/5	28		54		Catch/2
25/5	25/5	3		57		Normal
28/5	28/5	16		73		Normal
29/5	29/5	29		102		Normal
30/5	30/5	1		103		Normal
31/5	31/5	6		109		Normal

Fig. 13.6. Example of output from the option 'view data between given dates'. The output is for the Wye trap in 1989 between 1 January and 31 July for *Phorodon humuli*. The first two columns show the start date and end date for each trapping period in which the species was caught. The next four columns show the total number of aphids caught (males plus females), the number of males caught, and the cumulative totals of all aphids and just males. The final column shows the status of the trap; normal indicates that the trap was running correctly for the entire trapping period, Catch/2 indicates that the catch was halved, either by restricting the trap or by dividing the catch in the laboratory.

1. Information delivery (bulletin production)
2. Information analysis
3. Information interpretation (advisory (expert) systems)

INFORMATION DELIVERY – BULLETIN PRODUCTION

An aphid bulletin, providing information on the current numbers of particular pest species in suction trap samples, has been produced by RIS for each week during the crop season since 1968. The bulletin is a summary of the counts of 21 agriculturally important aphid species, or species groups, for the 14 traps currently operating. FLYPAST now enables fully automated production of the bulletin.

Summaries of data for longer periods and for previous years can also be produced simply by selecting the appropriate option from within the program.

INFORMATION ANALYSIS

The principal purpose of FLYPAST is to improve access to trap data. This has been achieved by developing a suite of programs to meet the needs identified in the initial specification. Each of these is customized by first selecting the year, site, and species of interest from multiple choice menus. Then the user can carry out the analyses shown in Table 13.2 by selecting the appropriate menu option.

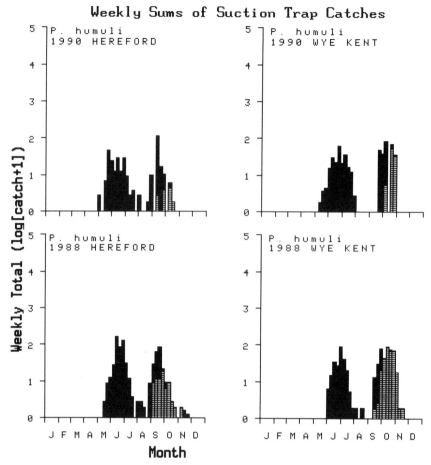

Fig. 13.7. Example of output from the option 'graph weekly totals' comparing the catches at two sites, Wye and Hereford, in 1988 and 1990 for *P. humuli*. This option can be used to show any differences that may occur between the sites or between years (males and females ▮ , males ≣).

INFORMATION INTERPRETATION – ADVISORY (EXPERT) SYSTEMS

The advisory systems in FLYPAST are designed to facilitate the use and interpretation of models using suction trap data. To run these models data are obtained from two sources: from the answers given by the user concerning site-specific features, and from the database. The output is in the form of a forecast or recommendation. So far, two systems have been developed, one for potato virus Y in home-saved seed potatoes, and the second for *Aphis fabae* Scop. in field beans.

Potato virus Y – The potato virus Y (PVY) system is designed to assist growers in deciding whether to save tubers for use as seed in the following season. For the use of home-saved seed to be worthwhile, the farmer must be sure that there is less than 5% infection with aphid-borne viruses, such as PVY, which could cause a significant reduction in yield. The system described here is based on the relationship identified previously

FLYPAST
Accessed on 22/08/91 at 13:55:45

Site WYE KENT Year: 1989		
Catches recorded between 1 JAN and 31 DEC		
Total number of days = 365		
Species: P. humuli		
	Males & Females	Males
Total Aphids	862.00	31.00
Average per day	2.36	0.08
Number of catches	80.00	17.00
Average per catch	10.78	1.82
Maximum in catch	62.00	4.00

Fig. 13.8. Example of output from 'statistics between set dates' option. The totals and means for the chosen period are for the Wye trap in 1989 and for *P. humuli.*

by Harrington *et al.* (1986) between the numbers of vector aphids caught in the suction traps and the level of infection in nearby crops. The model gives an estimate, before harvest, of the level of PVY which can be expected in a particular variety in a given area. The grower can then make an early decision on whether to buy new seed if high levels are predicted, to have tubers tested by ADAS*, if there is still uncertainty about the best option to choose, or, to retain some tubers for seed if PVY levels are expected to be below a threshold value.

In working through the system, the user is prompted for information about the particular type of potato crop of concern: its location, maturity class, cultivar, date of crop emergence, and the grade of seed planted. This information is combined with data from the database, to calculate the 'infectivity pressure' on the crop. The estimated incidence of virus in the crop is then presented in graphical form in Chapter 6, this volume (see p. 92). The left side of the diagram shows the weekly catches of the vector species from the local trap in June and July for the year in question, together with a long-term average. The right side of the diagram shows an estimate of the level of infection in the crop, which can be used by the grower to make a decision.

Aphis fabae **decision support system** — The decision support system for forecasting *Aphis fabae* levels in spring-sown beans is based on a survey of aphid populations on different host plants and the numbers of aphids migrating between them. This system, which uses the relationships between egg counts and suction trap samples, and the infestation level of *Aphis fabae* on spring-sown beans (Way *et al.*, 1977, 1981) provides advice based on an economic analysis of control options. This is described in more detail by Knight and Cammell (in preparation).

---

* ADAS = Agricultural Development and Advisory Service.

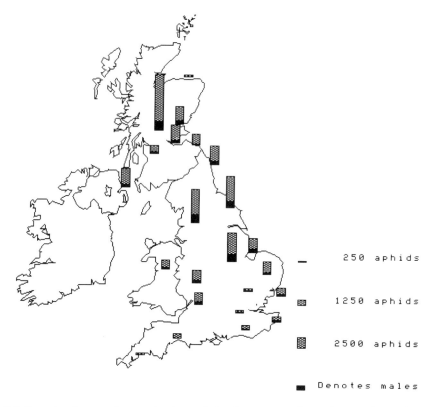

TRAP CATCHES FOR 1988 R. padi
SEPTEMBER 1   to DECEMBER 31

—   250 aphids

▨   1250 aphids

▦   2500 aphids

▬   Denotes males

Fig. 13.9. Example of output from option 'map all sites between set dates' showing the total number of *Rhopalosiphum padi* caught in 1988 for all traps operating that year. This display allows easy comparison of the distribution of catches over the United Kingdom.

Utilities

FLYPAST contains a number of programs to assist in maintenance of the system. The functions performed include:

- Editing of database files should corrections or alterations be necessary.
- Compression of large database files for storage and backup, reducing their size by about 80%. Expansion of compressed files before use.
- 'Back up' of the complete system, all database files, or just the aphid database files. This simplifies the backup procedure and increases the likelihood of it being done, reducing the chances of any serious loss of data.

PLANNING A DATABASE PROJECT

Planning, developing and implementing a database system follows steps similar to those required in any software development as discussed earlier in Chapter 6. These steps are briefly discussed here to provide a check-list of points that should be considered when planning any database development.

Project evaluation – The first step in the development of a database system is to decide whether it is actually necessary. Listing the anticipated potential benefits and possible problems with the system is one way of making the evaluation. Some of the benefits that may be anticipated from a system may not be realized, and others may not be appropriate in a particular situation. It is sometimes perceived that a database system will reduce the amount of time spent in maintaining files of information, but this is not necessarily the case. The extra capabilities offered by a computer database system over a paper filing system can easily generate enough extra work to occupy any time saved!

User requirements – The potential users of a system must be identified, and their requirements for the system determined. This requires an analysis of any current procedures that the system will replace or complement, as well as the identification of potential improvements that the system could deliver. Example or demonstration programs can be useful at this stage in showing the users the kind of facilities that the system can provide, particularly if they have not used database systems previously.

System specification – Having decided with the users what the system will be expected to do, the components of the system must be specified in more detail. For the users, the most important part of the specification is what they will see on screen, called the user interface, and what they will be able to do with the system. For the programmer, more technical detail is necessary, concerning the language or package that will be used to develop the system, the structure of the database, and the various program procedures that will be required. Included in a system specification may be some indication of what future developments might be required, and how they will be incorporated into the system.

Writing a precise system specification is easier for systems such as accounting packages where the procedures that the system will be required to perform are already well defined. In pest management, the decision processes that a database system may be required to support and improve are often less well defined. A major component of the project can sometimes be the development of methods for using the data, which can require a substantial research effort not envisaged at the start of the project. A radical rethink of the whole process of data collecting, analysis and use will often be desirable. This might best be carried out in a structured workshop using some of the primary decision tools discussed earlier in this book and involving the participation of potential users, relevant scientists and software developers.

Software development – Once the system specification has been written, the programming can be undertaken. If the specification has been properly written, the programming can be done by a programmer who has no particular knowledge of the field of interest of the end-user. If possible, it is preferable to have the programming done by professional programmers, with a systems analyst in overall charge of the project. However, funding constraints and the absence of a precise specification can often preclude this. Early on in the programming, the development of prototypes for evaluation by users is a way of determining if the system will be acceptable.

Software testing – Software testing can take various forms, to check for the different types of problem that can occur. The programmer can test for system bugs, and the performance of the system, when given extreme or unusual data sets. The user must also be involved in testing the system operationally, to ensure it conforms to the specification and his or her requirements.

Implementation – Finally, the completed system can be implemented. Where the system is replacing existing paper-based procedures, the changeover should be phased, including a period of parallel operation. An important part of implementation is the provision of any training required, adequate documentation for day to day use, and technical documentation of the system, that will be required for any future adjustments and additions.

REFERENCES

Bardner, R., French, R.A. and Dupuch, M.J. (1981) Agricultural benefits of the Rothamsted Aphid Bulletin. *Rothamsted Experimental Station, Report for 1980 Part 2*, pp. 21–39.

Cammell, M.E., Tatchell, G.M. and Woiwod, I.P. (1989) Spatial pattern of abundance of the black bean aphid, *Aphis fabae*, in Britain. *Journal of Applied Ecology* 26, 463–472.

Harrington, R., Govier, D.A. and Gibson, R.W. (1986) Assessing the risk from potato virus Y in seed saved from potato crops grown in England. *Aspects of Applied Biology* 13, 319–323.

Harrington, R., Tatchell, G.M. and Bale, J.S. (1991) Weather, life cycle strategy and spring populations of aphids. *Acta Entomologica et Phytopathologica Hungarica* 25, 423–432.

Knight, J.D., Tatchell, G.M., Norton, G.A. and Harrington, R. (1992) FLYPAST. An information management system for the Rothamsted aphid database to aid pest control research and advice. *Crop Protection* 11, 419–426.

Odiyo, P. (1990) Progress and developments in forecasting outbreaks of the African armyworm, a migrant moth. *Philosophical Transactions of the Royal Society of London, Series B* 328, 555–569.

Tatchell, G.M. (1985) Aphid-control advice to farmers and the use of aphid-monitoring data. *Crop Protection* 4, 39–50.

Tatchell, G.M. (1991) Monitoring and forecasting aphid problems. In: Peters, D.C., Webster, J.A. and Chlouber, C.S. (eds) *Aphid–Plant Interactions: Populations to Molecules*. Oklahoma Agricultural Experimental Station, Miscellaneous Publication No. 132, pp. 215–231.

Taylor, L.R. (1973) Monitoring surveying for migrant insect pests. *Outlook on Agriculture* 7, 109–116.

Taylor, L.R. (1977) Migration and the spatial dynamics of an aphid, *Myzus persicae*. *Journal of Animal Ecology* 46, 411–423.

Way, M.J., Cammell, M.E., Alford, D.V., Gould, H.J., Graham, C.W., Lane, A., Light, W.I. St.G., Rayner, J.M., Heathcote, G.D., Fletcher, K.E. and Seal, K. (1977) Use of forecasting in chemical control of black bean aphid *Aphis fabae* Scop., on spring-sown field beans, *Vicia faba* L. *Plant Pathology* 26, 1–7.

Way, M.J., Cammell, M.E., Taylor, L.R. and Woiwod, I.P. (1981) The use of egg counts and suction trap samples to forecast the infestation of spring-sown field beans, *Vicia faba*, by the black bean aphid, *Aphis fabae*. *Annals of Applied Biology* 98, 21–34.

Woiwod, I.P., Tatchell, G.M. and Barrett, A.M. (1984) A system for the rapid collection, analysis and dissemination of aphid-monitoring data from suction traps. *Crop Protection* 3, 273–288.

14 Geographic Information Systems

D.L. JOHNSON

Research Station, Agriculture Canada, Lethbridge, Alberta T1J 4B1, Canada

INTRODUCTION

Many of the decision tools discussed in other chapters in this book, such as historical and seasonal profiles, emphasize the temporal dimensions of pest problems. While the database systems used to investigate migratory pest problems (Chapter 13) incorporate both temporal and spatial dimensions of pest problems, the emphasis again is still very much on temporal aspects. Geographic information systems (GIS), as the name implies, are tools that are specifically designed to handle the spatial aspects of agricultural and forest pest management problems. Following a brief introduction to GIS, the general purposes for which it can be used, and the types of systems available, two specific uses for which GIS has been used are described. In both cases, the pest problem concerns grasshoppers in the Canadian prairies. In the first example, GIS is used to investigate surveying and forecasting of grasshoppers, while in the second example, it is used to examine the relationship between grasshopper outbreaks and soil class.

BACKGROUND TO GIS

GIS technology is used for three main purposes: for the organization and storage of data, for monitoring and management of resources or activities, and as a modelling and research tool. GIS applications are also being developed in a wide range of disciplines and businesses, such as urban and regional planning, engineering, forestry, earth sciences, marketing, political analysis, defence, environmental protection, real estate marketing, and studies of global climate change. Pest management applications are more common in forest pest management than in agricultural systems (Coulson *et al.*, 1988; Gage *et al.*, 1988; Liebhold and Elkinton, 1989). Lucid introductions to the techniques common to all these areas of GIS application are given by Burrough (1988) and Aronoff (1989).

Much of the recent attention given to large-scale spatial dynamics has been related to insect migration as a factor in synoptic pest studies (Fisher and Greenbank, 1979;

Taylor, 1986). However, even pests with limited dispersal and whose distribution and abundance are effected primarily by local conditions, can also be usefully studied in a spatial context when they occur over a large area or are affected by geographic variables such as weather and cropping geography.

Although there are many varieties of GIS, they all utilize computer-based methods of storing, summarizing and analysing data with reference to its geographic position. The more versatile GIS include input methods such as digitizing software, dBASE linkages and peripheral devices and software for jobs such as importing remote sensing data. Such systems are capable of handling large quantities of data and conducting calculations and comparisons involving multiple thematic maps and associated tables of attributes. Some GIS are designed for a specific purpose, such as climatic matching (see CLIMEX in Chapter 6), while others are generic tools that must be adapted to each specific problem by the user.

The principal feature of a GIS is that it is designed to incorporate information on position and topological relationships; for instance, are components neighbours, or are they nested? Typical spatial data have one, two or three dimensions — points, lines or areas — with associated information on measured or predicted attributes, such as temperature, elevation or crop stage. The areas can be manipulated using vectors (lines with nodes, endpoints and direction), raster (positions on a grid), or quadtrees (variable-size squares, based on the breakdown of an image, according to the level of detail and complexity).

The GIS software available usually provides at least two of these methods for doing calculations and, in most cases, simple tasks can be performed equally well via several methods. For example, the extent of severe levels of aphid infestation can be found by numerically determining the area of the polygons outlined by the boundaries (vectors) of the severe class on a map, or the same answer could be calculated by summing the number of small squares, such as pixels of a computer image, falling within the regions of interest. The next step might be to overlay the aphid map with maps of crop stage and recent weather, to calculate a new map showing the intersections where pest management action may be necessary, and this task could be performed using vector, raster, or quadtree methodology.

Most GIS software is also capable of some form of modelling that allows the user to build equations in which maps, such as pest density, recent rainfall, or crop stage, can be used as independent variables and can be combined according to specified relationships to allow calculation of a predicted map. This type of modelling serves three purposes:

1. It allows more sophisticated summaries of the data to be made, using algebraic operations; for instance, a map of practical crop value can be calculated from a map of yield data, adjusted according to the distance from markets.
2. A GIS model can 'look back' and be used to develop ideas about how geographic variables are related; for example, one could overlay and calculate the degree of correspondence of elevation and the incidence of a certain crop disease.
3. On the basis of hypotheses generated from **1** and **2** above, predictive GIS models can be developed, as in the case of forecasting the degree and distribution of pest damage.

In the last decade, the advent of faster, smaller computers and more efficiently written software has revolutionized GIS applications. There are scores of powerful GIS packages available (Parker, 1989). Most of the older software programs ran on mid-size to large mainframes, and many of these have been rewritten and adapted for PC DOS and OS/2 environments. Now many GIS packages are specifically designed for work stations utilizing UNIX or similar operating systems.

Not all packages perform the same tasks, and sometimes they perform best when used together. To illustrate the range available, let us consider the systems that are currently in use at the Lethbridge Research Station. (The list does not include numerous locally written computer programs, designed for specific tasks.)

GRASS – (Geographic Resources Analysis System, US Army Corp of Engineers) – a public domain computer program that is especially useful for work station users who want access to code, in order to modify some of the numerous raster-based GIS and image analysis procedures it performs. It is written in the C programming language and runs in the UNIX operating system

IDRISI – (a grid-based geographic analysis system from Clark University, MA, USA) – a DOS raster-based GIS that performs well for teaching and simple modelling

PAMAP – (a spatial analytical GIS from Pamap, Victoria, Canada) – a versatile system that offers sophisticated and flexible analytical tools and output

SPANS – (Spatial Analysis System, Interra Tydac, Ottawa, Canada) a system that offers rapid modelling with its quadtree structure, and access to many map projections and data conversions

Arc/Info – (ESRI, Redlands CA, USA) – used for a soils information system

Prices of these GIS systems vary considerably, from inexpensive (but surprisingly powerful) teaching packages, such as IDRISI for under US$300, to a more typical price of US$12,000 and up. The identification of users' requirements and the factors involved in selection of a system are covered by periodic surveys of GIS software needs and products (e.g., see Guptil, 1989). Parker (1989) presents the results of a survey of GIS users, in which he compares the basic attributes of 62 GIS packages.

The initial, and still the principal, application of GIS at the Lethbridge Research Station is the production of a grasshopper forecast; this case serves as a straightforward example of a pest management application that uses the simple GIS model types previously discussed.

GRASSHOPPERS IN THE CANADIAN PRAIRIES

Historically, the chief pests of cereal crops and grassland in the Canadian prairies are grasshoppers (Orthoptera: Acrididae), which periodically cause severe damage to crops and rangeland over a large geographic area. The costs of crop losses from grasshopper damage and of the resulting control measures can be extensive. For example, in 1985–86, insecticides used for grasshopper control cost Alberta farmers $12 million. Application costs totalled an additional $12 million, and crop losses in spite of control actions during the height of the infestation probably exceeded an additional $40 million. One of the main tools required to prevent heavy crop loss during rapidly growing infestations is advance warning of where best to allocate additional surveying effort, insecticide inventories, and training and extension services.

The outbreaks are typical examples of large-scale spatial dynamics that are affected by local conditions. Infestations begin with an expansion of resident populations in isolated locations, and appear to spread out and change geographical position over subsequent years as the rates of reproduction and survival change. Factors affecting the numerical fluctuations in grasshopper populations are usually variables that have both spatial and temporal characteristics (e.g., weather, soil, crop type), and can be mapped and incorporated into a GIS. One forecasting example and one historical analysis of grasshopper spatial dynamics are presented here, to illustrate some typical applications of GIS.

Surveying and Forecasting the Grasshopper Threat to Crops

Advance warning of changes in the geographical pattern and severity of grasshopper outbreaks is required to plan control measures. To meet this need, surveys of grasshopper abundance in Alberta have been conducted since the 1920s. One of the first practical applications of GIS to the problem of improving the forecast was the assessment of survey sampling methodology (Johnson, 1989a).

The method of sampling was standardized in 1932, and since 1987 has provided the basis for using GIS to produce the annual Alberta grasshopper forecast. Agricultural field personnel conduct detailed surveys each year in early August when most grasshoppers are in the adult stage. At each location, trained surveyors record the vegetation type and grasshopper counts in 100 m transects along the roadside, and in similar 100 m walks through the adjacent field. The data are collected from about 1500 to 2000 sites per year, and a summary of the grasshopper problem in the current year is produced. This summary is used to forecast the risk of crop damage for the coming year (Fig. 14.1, Table 14.1).

Until recently, the grasshopper forecast consisted of a map of the unadjusted adult population density (Johnson and Andrews, 1986), based on a system of combining roadside and field counts to produce risk ratings (Riegert, 1968; Smith and Holmes, 1977). This type of map, as Edwards (1962) pointed out, is 'not so much a forecast of what can be expected to happen in the future as a record of what has happened in the past.' In this comment he anticipated one of the ways in which GIS would be used to develop models based on past behaviour so that they could be used to make true, short-term forecasts.

Grasshopper recruitment and timing, and therefore the expected threat to crops in the following season, has been predicted by adjusting adult population density by heat accumulation (Randell and More, 1974; Hardman and Mukerji, 1982). Since these models were site models that do not include geography, simple maps of expected distributions were produced by these models, using regional averages.

In 1987, GIS technology was adapted to summarize grasshopper population dynamics, without reducing extensive data sets to district averages (Johnson and Worobec, 1988; Johnson, 1989a). Map overlay analysis and modelling was used to adjust population maps by maps of monthly rainfall and hours of sunshine in Alberta.

To assess the dependence of grasshopper populations on rainfall and sunshine hours, maps of population growth rate were produced by dividing each map of grasshopper density for one year by the same map for the preceding year. Regression analysis was used to produce a first model. The independent variable, population density in year

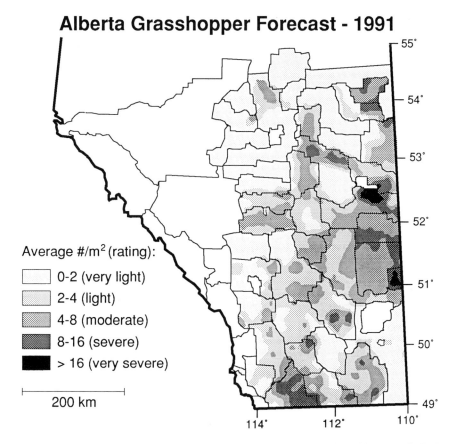

Fig. 14.1. The 1991 grasshopper forecast, generated with a PC-based GIS, and released four months before the field season. (The PC graphics package Mirage was used to generate the final map.)

Table 14.1. GIS-generated table of area of risk categories, based on grasshopper breeding density. Regions rated moderate and higher will require control measures on significant numbers of farms.

	1990 forecast		1991 forecast	
Legend	Area (%)	Area (km²)	Area (%)	Area (km²)
Very light	39.9	74.610	42.2	84.874
Light	41.5	77.600	25.3	50.663
Moderate	10.7	20.010	24.7	49.490
Severe	6.1	11.410	7.0	14.023
Very severe	1.8	3.370	0.8	1.516
Total	100.0	187.000	100.0	200.566

(*t* + 1) was fitted as a linear function of the population in year (*t*), hours of sunshine in August (*t*), and rainfall during August (*t*), June (*t* + 1) and July (*t* + 1).

The fundamental relationships apparent in the regressions and correlations were incorporated into equations describing the hypothetical effects of increased rainfall in decreasing population growth, and these were used to form the first simple GIS-based forecast. Each year, new data from weather stations and grasshopper surveys are used to update and improve the predictions.

Assessment of Survey Methodology

GIS methods have also been used to explore the potential for making survey methods more cost effective. By producing estimates of spatial autocorrelation of grasshopper population density, the relationship of roadside counts to field counts, and therefore the most efficient sampling method, could be determined.

The total database of grasshopper survey counts for the years 1978 to 1987 were used. Each record consists of the legal land description of the survey site (meridian, section, township, range), the two grasshopper population density estimates (field and roadside grasshopper counts per square metre), and the crop type. The analyses were undertaken in the four parts:

1. Correlation of roadside and field counts, to determine the degree to which the paired values of the two sampling methods covaried.
2. Spatial autocorrelation, to characterize the degree of geographical dependence of grasshopper densities, as determined by roadside and field counts.
3. Contour mapping of population density, using the results of the previous analysis to provide estimates of weighting function parameters required for estimating the population density contours. All contouring of the population data was accomplished with SPANS, but applications to generate smoothed maps from point data are readily available using other GIS packages (such as GRASS), or programs written for this purpose. After the grasshopper population maps for each year were contoured, the GIS was used to calculate the area that fell in each of seven density classes of grasshoppers per square metre. Cross-tabulations of the areas in the density categories were produced to construct a computational overlay of the roadside count and field count maps, by summing the areas of the 49 possible unique intersections.
4. Statistical prediction of field counts from roadside counts and crop type, using simple linear regression. The regressions were fitted to all the survey data from the first seven years of the study, so that they could be applied to the final three years as a test. This simple statistical model was used to predict the field counts for 1985 to 1987, the most recent years in which the forecast was prepared.

The results showed that while the roadside population distribution maps indicate geographic trends similar to those of the maps based on field data, high roadside counts.tend to overestimate field counts. For example, in 1985, 43% of the surveyed area had average densities greater than six adult grasshoppers per square metre in roadsides in late summer, compared with 30.8% for field counts. Nevertheless, for practical survey purposes, there is extremely good correspondence in the geographic distribution and abundance of grasshoppers predicted by the two methods.

Historical Analysis of the Spatial Relationship of Grasshopper Outbreaks to Soil Classification

Two spatial modelling methods, based on analysis of either area or point data, were used to test two hypotheses concerning the observation that grasshopper abundance in southern Alberta over the last ten years has been noticeably greater in certain soil zones than in others. The first hypothesis attributes this to differences in the physical properties of the soil, such as soil surface texture (Isely, 1937). The most likely mechanisms would be the effects of soil drainage, friability, moisture capacity, and thermal conductivity on grasshopper egg-laying, embryo development, and hatching success (Hewitt, 1985; Johnson *et al.*, 1986). A non-compacted loam or sandy soil would presumably allow a greater reproductive rate than clay, and would promote the rapid population growth that results in outbreaks.

A second hypothesis ascribes the greater abundance of grasshoppers in certain soil zones to geographic covariables and not to the particular physical qualities of the soil. For example, grasshopper populations may respond more to the vegetation type, farming practices, and weather in the different soil zones than to the attributes (e.g. sandiness) of the soils themselves.

These hypotheses were tested, using GIS, in two simple ways: by examination and analysis of map area intersections, and through analysis of point data. For area analysis, the observations on all variables (in this case, insect population samples and soil survey classes) are converted into maps. Tables describing the intersections of the map variables are constructed by overlaying the maps to find all unique conditions, calculate their areas, and determine the correlation or dependence among maps.

The second, and more conventional, method is based on analysis of the population sample data recorded at survey locations. With this second, point modelling method, the soil attributes must be determined and recorded for each survey site and used as independent variables in the regression model.

Grasshopper population data were obtained from the annual Alberta grasshopper survey, described above. A vector file delineating the soil classification areas was used to produce a map of soil polygons, and these were recoloured to produce maps of soil type and soil texture. To construct a table of map intersections for each of the ten years, the following maps were included in the GIS overlay:

1. Grasshopper density in the year of interest
2. Grasshopper density the year before
3. Soil type
4. Soil texture

The grasshopper density categories (the dependent variable, number per square metre) were log-transformed, and a weighted analysis of covariance was fitted to the entire ten-year data set, with the previous year's grasshopper density as the concomitant variable and intersection area as the weight. Year, soil type and soil texture were the main effects in the model (Table 14.2).

The points modelling approach used the same statistical model, but the analysis was unweighted, because raw survey data points and not map areas were used. The matrix was constructed by finding the soil attributes for each of the 12,247 survey sites that fell within the soils study area. SPANS was used to reference the soil maps, determine

Table 14.2. Analysis of covariance of ln(grasshopper density), as a function of the previous year's density, the soil type and the soil texture. The results of both area modelling and points modelling are shown. (DF = degrees of freedom; F = F statistic; P = probability.)

Source	Area modelling			Points modelling		
	DF	F	P	DF	F	P
Year	9	32.70	<0.001	9	20.19	<0.001
Soil type	5	8.48	<0.001	5	5.04	<0.001
Soil texture	5	1.66	0.165	5	0.55	0.734
Type × texture	12	1.37	0.189	12	1.52	0.130
Type × year	45	5.33	<0.001	45	5.54	<0.001
Texture × year	45	3.26	<0.001	45	3.81	<0.001
Type × text. × yr	108	3.07	<0.001	97	3.03	<0.001
Last year's popn	1	1.610	<0.001	1	1.149	<0.001
		SS (III)			SS (III)	
Residual	8.414	500.553		10.949	4.286	
Total	8.644	1154.024		11.168	7.272	

the soil type and texture, and append these values to the vector of grasshopper survey counts (observed number per square metre).

The results showed that both models accounted for significant proportions of the total variance of log-transformed densities (56% by area modelling and 41% by points modelling). Differences among years were highly significant in both cases, the previous year's population density accounting for more of the total variance in present population density than any other factor.

Both analyses also suggest that soil type is a highly significant factor ($P < 0.001$) but that the interaction — soil type × texture — are not ($P > 0.1$). This implies that although geographical distribution of grasshoppers in Alberta does indeed vary with soil type, it does not vary significantly with the sand, silt or clay content of the soil, either across the study area as a whole or within the major soil zones.

This comparative analysis also indicates that both area modelling and points modelling can be used reliably to compare variables over large geographical areas. The example in this case relied on detailed maps with large numbers of observations and unique conditions. However, in most cases, area modelling is faster than points modelling methods because of the considerable reduction in data handling.

FUTURE GIS DEVELOPMENT

Present research at Lethbridge Research Station concerns refining our understanding of the geographic distributions of insects, assessing the accuracy of the forecasts with overlays of predicted and observed maps, conducting studies on the geographic patterns of insecticide application, and consideration of applications of GIS to other agricultural problems (such as forecasting expected crop yield).

The insecticide sales GIS database contains information on over 12,000 sales, and this study database has been applied to both insect pest management and to the problem of predicting regions where there is a likelihood of impact of insecticide spraying on threatened wildlife species.

Other agricultural GIS projects include:

- A crop monitoring and forecasting GIS for rice production and pest management in an irrigation district of northern Malaysia — funded by Canadian International Development Agency, and currently in progress.
- An integrated crop—pest model for the Canadian prairies, with an attempt to tie in remotely sensed data, as has been utilized in some locust forecasting — currently being planned.

REFERENCES AND BIBLIOGRAPHY

Agriculture Canada Expert Committee on Soil Survey (1987) *The Canadian System of Soil Classification*, Second edition. Agriculture Canada Publication No. 1646.

Aronoff, S. (1989) *Geographic Information Systems: A Management Perspective*. WDL Publications, Ottawa, Canada.

Burrough, P.A (1988) *Principles of Geographical Information Systems for Land Resources Assessment*. Clarendon Press, Oxford.

Cherlet, M. (1990). Remote sensing in locust control. *Proceedings of a Symposium on Practical Applications of Agrometeorology in Plant Protection, Firenze, Italy, 4–6 December, 1990.*

Cliff, A.D. and Ord, J.K. (1981) *Spatial Processes Models and Applications*. Pion Ltd, London, 266 pp.

Coulson, R.N., Graham, L.A. and Lovelady, C.N. (1988) Intelligent geographic information system for predicting the distribution, abundance and location of southern pine beetle infestations in forest landscapes. In: Payne, T.L. and Saarenmaa, H. (eds) *Integrated Control of Scolytid Bark Beetles*. Blacksburg, Va., pp. 283–294.

Diggle, P.J. (1983) *Statistical Analysis of Spatial Point Patterns*. Academic Press, London..

Edwards, R.L. (1962) A critical appraisal of grasshopper forecast maps in Saskatchewan, 1936–1958. *Journal of Economic Entomology* 55(3), 288–292.

Fisher, R.A. and Greenbank, D.O. (1979) A case study of research into insect movement: spruce budworm in New Brunswick. In: Rabb, R.L. and Kennedy, G.G. (eds) *Movement of Highly Mobile Insects: Concepts and Methodology in Research*. North Carolina State University, Raleigh, NC., pp. 220–229.

Gage, S.H., Simmons, G.A. and Parks, B.O. (1988) Computer based geographic information systems for regional decisions in pest management. *Ext. Bull. Coop. Extension Services, Michigan State University* No. 2142, pp. 52–58.

Gilbert, N.E. (1972) *Biometrical Interpretation*. Clarendon Press, Oxford.

Grace, B.D. and Johnson, D.L. (1985) The drought of 1984 in southern Alberta: its severity and effects. *Canadian Water Research Journal* 10, 28–38.

Guptil, S.C. (1989) Evaluating geographic information systems technology. *Photogrammetric Engineering and Remote Sensing* 55, 1583–1587.

Hardman, J.M. and Mukerji, M.K. (1982) A model simulating the population dynamics of the grasshoppers (Acrididae) *Melanoplus sanguinipes* (Fabr.), *M. packardii* Scudder, and *Camnula pellucida* (Scudder). *Research in Population Ecology* 24(2), 276–301.

Hewitt, G.B. (1985) *Review of the Factors Affecting Fecundity, Oviposition, and Egg Survival of Grasshoppers in North America*. USDA. ARS-36.

Isely, F.B. (1937) Seasonal succession, soil relations, numbers, and regional distribution of northeastern Texas acridians. *Ecological Monographs* 7, 317–344.

Isely, F.B. (1938) The relations of Texas acrididae to plants and soils. *Ecological Monographs* 8, 551–604.

Johnson, D.L (1989a) Spatial autocorrelation, spatial modelling, and improvements in grasshopper survey methodology. *The Canadian Entomologist* 121, 579–588.

Johnson, D.L (1989b) Spatial analysis of the relationship of grasshopper outbreaks to soil classification. In: *Estimation and Analysis of Insect Populations. Lecture Notes in Statistics* 55, 347–359. Springer-Verlag, New York.

Johnson, D.L. and Andrews, R.C. (1986) *1986 Grasshopper Forecast.* 35 × 62 cm color map sheet (1:2,000,000), Alberta Bureau of Surveying and Mapping, Edmonton, Alberta.

Johnson, D.L. and Worobec, A. (1988) Spatial and temporal computer analysis of insects and weather: grasshoppers and rainfall in Alberta. *Memoirs of the Entomological Society of Canada* 146, 33–48.

Johnson, D.L., Andrews, R.C., Dolinski, M.G. and Jones, J.W. (1986) High numbers but low reproduction of grasshoppers in 1985. *Canadian Agricultural Insect Pest Revue* 63, 8–10.

Kemp, W.P., Kalaris, T.M. and Quimby, W.F (1989) Rangeland grasshopper (Orthoptera: Acrididae) spatial variability: macroscale population assessment. *Journal of Economic Entomology* 82, 1270–1276.

Liebhold, A.M. and Elkinton, J.S. (1989) Characterizing spatial patterns of gypsy moth regional defoliation. *Forest Science* 35, 557–568.

Parker, H.D. (1989) GIS software 1989: a survey and commentary. *Photogrammetric Engineering and Remote Sensing* 55, 1589–1591.

Randell, R.L., and More, R.B. (1974) *Modelling: VI. Population Dynamics of a Species of Grasshopper.* Canadian Committee for the International Biological Programme, Matador Project, Technical Report No. 59, 118 pp.

Riegert, P.W. (1968) A history of grasshopper abundance surveys and forecasts of outbreaks in Saskatchewan. *Memoirs of the Entomological Society of Canada* 52, 1–99.

Samet, H. (1984) The quadtree and related hierarchical data structures. *Computing Surveys* 16, 187–260.

SAS Institute (1982) *SAS User's Guide: Statistics.* SAS Institute, Cary, NC.

Smith, D.S. and Holmes, N.D. (1977) The distribution and abundance of adult grasshoppers (Acrididae) in crops in Alberta, 1918–1975. *The Canadian Entomologist* 109, 575–592.

Sokal, R.R. and Oden, N.L. (1978a) Spatial autocorrelation in biology. 1. Methodology. *Biological Journal of the Linnaean Society* 10, 199–228.

Sokal, R.R. and Oden, N.L. (1978b) Spatial autocorrelation in biology. 2. Some biological implications and four applications of evolutionary and ecological interest. *Biological Journal of the Linnaean Society* 10, 229–249.

Taylor, L.R. (1986) Synoptic dynamics, migration and the Rothamsted insect survey. *Journal of Animal Ecology* 55, 1-38.

15 Information Retrieval for Pest Management

C.J. HAMILTON

Library Services Centre, CAB International, Silwood Park, Buckhurst Road, Ascot SL5 7TA, UK

INTRODUCTION

Most, though not all, scientists recognize the sheer volume and diversity of current literature is such that they cannot keep up to date in anything other than the narrowest field of interest. For those involved in pest management, where the literature is not always clearly defined and where older data are often as or more important than new data, the problem is considerable.

This brief chapter aims to provide non-information specialists with an appreciation of the scale of the problem, and an introduction to the organizations, systems and technologies that can help them to overcome it.

More than 20,000 papers on crop protection are published annually. These are scattered among more than 2000 journals and around 1000 books, conference proceedings and other documents, and written in a total of some 40 languages.

Anyone involved in pest management research or decision making needs efficient access to the information contained − and sometimes buried − in this wealth of published literature. This is particularly important in biological control and integrated pest management, where questions such as 'where did the pest originate?' and 'what is known about its natural enemies?' are fundamental to a new programme.

Very few people have access to a large, comprehensive pest management library; even if they did, none could possibly keep up to date by simply scanning or reading the current primary literature. Similarly, exhaustive searches of the retrospective literature are rarely cost-effective and occasionally impossible without some sort of key to unlock the information. Printed abstracts and indexes have been providing pest management researchers with such a key for almost a century.

Abstract journals, like those produced by CAB International, allow the reader to keep informed not only of mainstream publications, but also those on the periphery and those in 'difficult' languages; they extract and present data from all of these in digestible form. Linked with document delivery services, abstracting services can help to provide equal information access to all, irrespective of geographical location.

Over the last 20 years most of the major abstracting services have computerized their journal production processes to make them more efficient. In the process, they have also created databases (computer-readable stores) of all the information in the printed journals. At the same time, a number of organizations began developing interactive computerized retrieval systems to deal with the large amount of information becoming available. There are now a number of highly sophisticated systems, each providing international access to a wide range of databases, including a number of particular relevance to pest management.

Abstracts are stored in such systems so as to allow the computer to locate and display those containing specific words, index terms, authors' names, etc., either singly or in combination. In most cases searching is not restricted to index terms or keywords but can include any or all parts of the record, including the abstract. A particularly useful feature is the ability that most of these systems have to search for a wordstem, so that a single command can be used to retrieve, for example, references containing the terms 'parasite', 'parasitoid', 'parasitic', 'parasitism' or 'parasitology'.

In general, using such systems involves connecting a relatively simple electronic keyboard device or a personal computer (PC), via the normal telephone network, to a remote 'host' computer which then responds almost instantaneously to queries or commands typed by the user. Most host computers are able to serve a substantial number of simultaneous users, without any noticeable delay in response.

Online searching, as this process is known, is interactive, with the user gradually refining his or her search in response to the number and relevance of the records retrieved and displayed at his or her terminal. When satisfied with the result, the user can have some or all of the retrieved records printed out or downloaded to disk for further processing or study.

For users with a continuing interest, there is usually an automatic current-awareness option. A search is carried out in the usual way, and each step recorded in the host computer's memory. Subsequently, as each new batch of records is added to the database (usually monthly), the computer will automatically repeat the search on just those new records, and a printout of the selected abstracts is mailed to the user.

Obviously, there would be little appeal in a system that required its users to learn a complicated computer language, and so most of the systems use very similar, very simple commands. These are easily learned within an hour or so, although it obviously takes longer to become an adept and efficient searcher.

The most important advantage of online searching is that it enables a search that might have taken hours or even days to carry out manually to be completed in a matter of minutes. Results can be printed out by the computer, leaving the user's time free for interpreting and using the information retrieved. The whole process is relatively effortless, so that online searching is often described as exhaustive but not exhausting!

The costs of online searching comprise basically the following elements:

1. Equipment costs, starting from about £600 (US$1000) for a simple terminal or personal computer (PC) equipped with a modem (essential for communications).
2. Telecommunications charges, which vary according to the user's location.
3. Hourly access charges, which are paid to the system supplier and vary according to the database being used.
4. Print charges, which are also database specific.

As a guide, the total direct cost of searching most of the databases listed later in this chapter is currently about £60 (US$100) per hour, excluding prints. Although this may sound expensive, it should be remembered that a typical search may take 15 minutes or less, so that the overall cost is often relatively low.

The relatively recent but highly successful CD-ROM (Compact Disc Read-Only Memory) technology has brought similar power to the desktop, allowing information to be accessed whenever it is needed, 24 hours a day, 7 days a week, independently of telecommunications links and other external factors which may limit the availability of online databases.

CD-ROM is a well-established technology developed originally for the hi-fi industry. When used for digital data storage, a single 12 cm diameter CD-ROM disk can hold up to 600 megabytes of data, equivalent to 1500 floppy diskettes or 200,000 typewritten A4 pages. This vast capacity makes CD-ROM an excellent storage medium for large databases – more than 500,000 abstracts can be stored on a single disk, occupying less than a centimetre of shelf space.

The basic equipment needed to use CD-ROM databases is very simple: an IBM XT/ AT-compatible PC, preferably with a hard disk, and either linked to, or fitted with, a suitable CD-ROM drive. As a guide, a complete CD-ROM workstation with a good printer would cost around £1400 (US$2200). The number of databases available on CD-ROM is growing rapidly, and several are of direct interest to agricultural scientists, including AGRICOLA, BIOSIS and CAB ABSTRACTS. Many are sold on an annual lease basis rather like a journal subscription, although some (including CAB ABSTRACTS) may be purchased outright. Generally speaking, there are no usage charges, so users can make as much use of the databases as they wish without incurring any additional costs. Most CD-ROM search and retrieval software is both powerful and user-friendly – simple to learn yet powerful enough to give even novice users good results.

EXPLANATION OF A SAMPLE SEARCH

In the sample search shown (Fig. 15.1), the DIALOG computer system has been used to search the CAB ABSTRACTS database to find references on the biological control of the desert locust (*Schistocerca gregaria*).

A connection is first established with the DIALOG computer via the normal telephone network. This procedure (logging on) generally takes about 20 seconds, after which the computer prompts us with a '?' to indicate that it is ready to accept our commands, which are shown here in italics.

When searching, it is usually advisable to include variant terms: selecting 'biological control' will not necessarily retrieve all records relevant to this topic, so related terms (natural enemies, parasites, pathogens and predators) are also used.

DIALOG can search easily for words with a common stem so, for example, the command 'select parasit?' will retrieve all records including one or more of the terms parasite, parasites, parasitic, parasitism and parasitology.

Online searching uses the principles of Boolean logic, so that search terms can be combined using logical operators ('and', 'or', 'not'). In this particular example, we have also bracketed the search terms together, rather like an algebraic expression, for speed and convenience.

?select schistocerca gregaria and (biological control or natural enem? or parasit? or pathogen? or predat?)
     344 SCHISTOCERCA GREGARIA
  7027 BIOLOGICAL CONTROL
14834 NATURAL ENEM?
35617 PARASIT?
30956 PATHOGEN?
 8051 PREDAT?
S2   43 SCHISTOCERCA GREGARIA AND (BIOLOGICAL CONTROL OR
       NATURAL ENEM? OR PARASIT? OR PATHOGEN? OR PREDAT?)
?t1/7/1

1/7/1
 0955661 0E078-04099; 0J078-02071; 7E011-01836
   In vivo and in vitro assays for pathogenicity of wild-type and mutant strains of Metarhizium anisopliae for three insect species.
   Huxham, I. M.
   Department of Zoology, University of Glasgow, G12 8QQ, UK.
   Journal of Invertebrate Pathology 1989. 53 (2): 143-151 (28 ref.)
   Language: English
   Document Type: NP (Numbered Part)
   Status: REVISED
   Subfile: 0E (Review Applied Entomology, Ser. A); 0J (Review Applied Entomology, Ser. B); 7E (Biocontrol News and Information)
   A panel of assays is described which measures the ability of the entomogenous fungus Metarhizium anisopliae to grow (growth in vitro and LT50 in vivo) and the ability of the insect's haemocoelic immune system to respond (changes in haemocyte number and changes in the ability to respond to fungal pathogens by production of phenoloxidase in vitro). Using these assays, the relative pathogenicity of wild-types ARSEF 455, 488 and 551 and IHR 83-82 (QEC 410.0, 416,0, 420.0 and 403.0, resp.), the mutant strain QEC 410.20, and the recombinant strain QEC X22 were compared for their effects on Schistocerca gregaria, Periplaneta americana and Nilaparvata lugens. The methods outlined should facilitate identification of strains with traits which relate directly to pathogenicity within the host insect.

?logoff

Fig. 15.1. Sample online search.

DIALOG searches the entire database for records incorporating each term in isolation (a count is given for each), then combines them according to our specification; the final total here is a set (s2) of 43 records. We have then asked DIALOG to type just the most recent of these, in a format which includes the abstract and full bibliographic citation. More records could be typed out at our terminal if required, or we could enter a simple 'print' command to generate an 'offline' printout on DIALOG's own printer; this would then be airmailed to us within 24 hours.

Finally, we enter a 'logoff' command to disconnect from the computer. The total time for this particular search was just 0.03 hours (1.8 minutes).

SYSTEMS SUPPLIERS

The organizations listed below operate the host computers that provide access to the databases listed in the next section. They can provide information on the procedures to be followed in order to obtain a password and gain access to their particular systems, but queries relating to the content and use of any specific database are best directed to the relevant producer.

Intending users should remember that the physical location of the host computer is relatively unimportant; most of the systems listed here can be accessed at comparable cost from most parts of the world by means of established telephone links and/or private networks. The systems suppliers can provide information about these.

BRS Information Technologies, 1200 Route 7, Latham, New York 12100, USA.
CAN/OLE, National Research Council of Canada, Ottawa K1A 0S2, Canada.
DATA-STAR, Radio-Schweiz AG, Laupenstrasse 18A, CH-3008 Berne, Switzerland.
DIALOG Information Services, 3460 Hillview Avenue, Palo Alto, California 94304, USA.
DIMDI, PO Box 42 05 80, Weisshausstrasse 27, D-5000 Koln 41, Germany.
ESA-IRS, Via Galileo Galilei, I-00044 Frascati, Italy.
Human Resource Information Network (HRIN), College Park North, 9585 Valparaiso Court, Indianapolis, IN 46268, USA
JICST, Nihon Kagaku Gijutsu Joho Center, 5-2 Nagatacho 2-Chrome, Cihyoda-ku, Tokyo 100, Japan.
LEXIS, 9443 Springboro Pike, P.O. Box 933, Dayton, OH 45401-9964, USA
ORBIT Search Service, 8000 Westpark Dr., McLean, VA 22102, USA
STN International, c/o Chemical Abstracts Service, PO Box 3012, Columbus, OH 43210, USA.

DATABASES RELEVANT TO CROP PROTECTION

ABSTRACTS ON TROPICAL AGRICULTURE
Printed equivalent: Abstracts on Tropical Agriculture (ATA)
Time-span: 1975–
Size: 25,000+ records
Growth rate: 350/month

Producer: Royal Tropical Institute, Information and Documentation, Mauritskade 63, NL-1092 AD Amsterdam, The Netherlands.
Access systems: ORBIT
Coverage: Tropical and subtropical agriculture, with emphasis on extension-related information.
Comments: A useful but relatively small file.

AGRICOLA
Printed equivalent: Bibliography of Agriculture
Time-span: 1970–
Size: currently approx. 2,500,000 records
Growth rate: approx. 190,000 records per year
Producer: USDA, National Agricultural Library, Beltsville, Maryland 20705, USA.
Access systems: BRS, DIALOG
Coverage: All aspects of agriculture including applied entomology.
Comments: Useful database on general agriculture; includes a fairly high proportion of US extension literature; some abstracts but records mainly in the form of bibliographic citations only.

AGRIS
Printed equivalent: AGRINDEX
Time-span: 1975–
Size: currently approx. 1,057,000 records
Growth rate: approx. 120,000 records per year
Producer: AGRIS Coordinating Centre, FAO, via delle Terme di Caracalla, 00100 Rome, Italy.
Access systems: DIALOG, DIMDI, ESA-IRS
Coverage: All aspects of agricultural science and technology, including applied entomology.
Comments: Another general agricultural database; particularly good for extension literature and reports published in the LDCs (less developed countries); some abstracts but mainly bibliographic citations only.

AGROCHEMICALS HANDBOOK
Printed equivalent: Agrochemicals Handbook
Time-span: 1983–
Size: 750+ records
Growth rate: 20 per 6 months
Producer: Royal Society of Chemistry (RSC), Information Services, Thomas Graham House, Science Park, Milton Rd, Cambridge, CB4 4WF, UK.
Access systems: DATA-STAR, DIALOG (on CD-ROM as part of The Pesticides Disc)
Coverage: Contains factual data on approximately 750 substances that are active components of agrochemicals.
Comments: Data on the active components found in agrochemicals used worldwide. Includes chemical names, including synonyms and tradenames, CAS Registry Number, molecular formula, molecular weight, manufacturers' names, chemical and physical property, toxicity, mode of action, activity, health and safety, etc.

BIOSIS PREVIEWS
Printed equivalent: Biological Abstracts, Biological Abstracts RRM
Time-span: 1969–
Size: currently approx. 6,156,000 records
Growth rate: approx. 360,000 records per year
Producer: BIOSIS, 2100 Arch Street, Philadelphia, Pennsylvania 19103, USA.
Access systems: BRS, DATA-STAR, DIALOG, DIMDI, ESA-IRS, JICST
Coverage: All types of scientific and semi-popular literature on all aspects of life sciences, including pure and applied entomology.
Comments: A good database, particularly for pure entomology including taxonomy. Mainly bibliographic citations only, but abstracts are being included for some of the more recent literature. A relatively unusual feature is the hierarchical indexing which allows the use of broad search terms (e.g. insect family names) to retrieve records containing narrower terms (e.g. genera).

CAB ABSTRACTS
Printed equivalent: All CAB ABSTRACTS journals including Biocontrol News and Information, the Review of Agricultural Entomology etc
Time-span: 1973– (1972– for veterinary subjects)
Size: currently approx. 2,240,000 records
Growth rate: approx. 180,000 records per year
Producer: CAB International, Wallingford, Oxon OX10 8DE, UK.
Access systems: BRS, CAN/OLE (Canadian access only), DATA-STAR, DIALOG, DIMDI, ESA-IRS, JICST
Coverage: The world literature of agricultural science in the broadest sense, including applied entomology and taxonomy of economically important species.
Comments: The best database for applied entomology and apiculture. Easy to use, and almost all records incorporate an informative abstract in English.

CHEMICAL REGULATION REPORTER
Printed equivalent: As above (weekly)
Time-span: 1982–
Size: –
Growth rate: – (weekly updates)
Producer: The Bureau of National Affairs, Inc. (BNA), BNA ONLINE, 1231 25th St NW, Washington DC 20037, USA.
Access systems: Human Resource Information Network (HRIN); LEXIS
Coverage: Full-text reports of legislative and regulatory development affecting production and use of pesticides.
Comments: A useful full-text source, but not widely accessible.

CRIS/USDA
Printed equivalent: None
Time-span: Covers research projects which are currently active or terminated within previous two years.
Size: currently approx. 31,000 records
Growth rate: 0 (old project records are deleted after two years)

Producer: USDA, SEA/TIS, National Agricultural Library Building, Beltsville, Maryland 20705, USA.

Access systems: DIALOG

Coverage: Not a bibliographic database, but a collection of factual data on agriculture-related research projects. These are current and recently terminated projects sponsored or conducted by USDA research agencies and cooperating state institutions.

Comments: Very useful, particularly when used in conjunction with a retrospective search of one of the major bibliographic databases. Many of the projects detailed are not written up for publication for quite some time, if at all, so that the information contained may not be available elsewhere.

CSA LIFE SCIENCES COLLECTION

Printed equivalent: All CSA (Cambridge Scientific Abstracts) abstracting journals, including Entomology Abstracts

Time-span: 1978—

Size: currently approx. 1,058,000 records

Growth rate: approx. 120,000 records per year

Producer: Cambridge Scientific Abstracts, 7200 Wisconsin Avenue, Bethesda, Maryland 20814, USA.

Access systems: DATA-STAR, DIALOG, ESA-IRS

Coverage: Animal behaviour, biochemistry, ecology, entomology (pure and applied), genetics, immunology, microbiology, toxicology and virology, etc.

Comments: Still relatively small in comparison with the other databases which cover these topics, but good all-round basic coverage of entomology. Abstracts are included, but these tend to indicative rather than informative.

EUROPEAN DIRECTORY OF AGROCHEMICAL PRODUCTS

Printed equivalent: As above (4 vols)

Time-span: 1984—

Size: 24,000+ records

Growth rate: 1000 per 6 months

Producer: Royal Society of Chemistry (RSC), Information Services, Thomas Graham House, Science Park, Milton Rd, Cambridge CB4 4WF, UK.

Access systems: DATA-STAR, DIALOG (on CD-ROM as part of The Pesticides Disc)

Coverage: Factual data on a very wide range of European agrochemicals.

Comments: Useful source of factual data, including active ingredient proportions, application timing, toxicity, uses, etc.

PESTDOC

Printed equivalent: PESTDOC

Time-span: 1968—

Size: 300,000+ records

Growth rate: 1000 per 3 months

Producer: Derwent Publications Ltd, Derwent House, 14 Great Queen Street, London WC2B 5DF, UK

Access systems: ORBIT (Accessible only to subscribers to hard-copy)

Coverage: All aspects of pesticides including chemistry, toxicology, legislation, use and environmental effects.
Comments: Well-respected by users, but these are limited by the fact that access is only available to hard-copy subscribers.

PEST MANAGEMENT RESEARCH INFORMATION SYSTEM (PRIS)
Printed equivalent: Pesticide Information (quarterly)
Time-span: 1981–
Size: Varies (current data only)
Growth rate: –
Producer: Agriculture Canada, Scientific Information Retrieval Section, Central Experiment Farm
Access systems: CCINFOline (250 Main St, E, Hamilton, Ontario L8N 1H6, Canada). (Also on CD-ROM as part of CCINFOdisc)
Coverage: Factual/bibliographic data on pest control products from their introduction to Canada until they are registered for use in Canada.
Comments: Useful, but not as easily or widely accessible as others of its kind.

PHIND
Printed equivalent: Various, including AGROW World Crop Protection News
Time-span: 1980–
Size: –
Growth rate: Daily updates
Producer: PJB Publications Ltd, 18-20 Hill Rise, Richmond, Surrey, TW10 6UA, UK.
Access systems: BRS; DATA-STAR; DIALOG
Coverage: Full-text of all articles appearing in PJB Publications newsletters (see above).
Comments: Very useful full-text database, particularly for topical/industrial information on agrochemicals. Complements the more scientific slant of the major bibliographic databases.

PHYTOMED
Printed equivalent: Bibliographie der Pflanzenschutzliteratur (quarterly)
Time-span: 1965–
Size: 360,000 records
Growth rate: 4,000 per 3 months
Producer: Deutsche Biologische Bundesanstalt fur Land-und Forstwirtschaft, Dokumentationsstelle fur Phytomedizin, Konigin Luise Str. 19, D-1000 Berlin 33, Germany.
Access systems: DIMDI, STN International
Coverage: Phytomedicine and related field of agriculture, biology and chemistry.
Comments: Although quite a small database and lacking abstracts, it is unusual in that records are indexed in English, French and German.

16 Decision Tools for Biological Control

J.K. WAAGE[1] AND N.D. BARLOW[2]

[1]International Institute of Biological Control, Silwood Park, Buckhurst Road, Ascot SL5 7TA, UK: [2]Agresearch, New Zealand Pastoral Agricultural Research Institute Ltd, Ellesmere Junction Road, PO Box 60, Lincoln, New Zealand

INTRODUCTION

Most of the previous chapters in this book have concentrated on particular decision tools, with examples to illustrate their use in pest management. In this chapter, a different approach is adopted. Here, we concentrate on a particular form of pest management – biological control – and consider how decision tools might be of value in improving decision making in this specific area. First, the background to biological control is reviewed, as are the different approaches and the decisions that have to be made. It is in this context that the role of decision tools for biological control, particularly classical biological control, is discussed.

BACKGROUND TO BIOLOGICAL CONTROL

Many definitions have been applied to the term 'biological control'. We shall define biological control as the use of living organisms as pest control agents. Thus we shall not consider the use of natural products (pheromones, plant-derived insecticides and repellents), resistant plant varieties, or autocidal methods.

Less than a century ago, biological control was a logical option to which researchers would turn when faced with a new pest problem. Much precedent had been established for the introduction of exotic natural enemies to control pests, and for the regular mass production and release of natural enemies for the same purpose.

With the development of synthetic organic pesticides in the 1950s, interest in biological control decreased drastically. Synthetic pesticides proved more efficient than natural enemies and frequently had the effect of eliminating them from the crop, along with the pest. It was only when serious problems began to arise with pesticide use that biological control was re-examined.

In many crop systems, where pests had become resistant to chemicals, continued application of pesticides led to resurgence of the target pest, or of some secondary pests. In both cases, this could be traced to the preferential elimination of natural enemies relative to pests. Resurgence was particularly apparent with the use of organochlorine compounds, and against particular pest groups, such as Homoptera and Lepidoptera.

The growth of the concept of integrated pest management in the 1970s led to a rebirth of interest in biological control as one of the components of IPM. Since then, researchers have realized that it is probably a primary component of IPM, the role of natural enemies having been grossly underestimated in many crop systems.

APPROACHES TO BIOLOGICAL CONTROL

All pests have natural enemies. So why do we have pests? Clearly, the impact of natural enemies is often insufficient to prevent pests from causing losses to crop production. This may be simply because such losses occur at pest densities well below those which natural enemies can maintain in the field.

On the other hand, natural enemies may not be realizing their full potential to depress crop pest populations. Modern crop monocultures do not provide the range of alternative foods and microhabitats which many natural enemies depend on in more diverse, natural vegetation. Consequently, natural enemy numbers may be low in crops, or arrive too late during the cropping season.

Thus, a farmer, extension worker or scientist, interested in building biological control into a pest management programme, should first consider what natural enemies of the pest occur in the crop, and what might be limiting their effectiveness. This clearly may not be a trivial exercise.

Conservation of Natural Enemies

Nevertheless, once identified, a number of methods might be used to conserve these important natural enemies. This might be achieved through modifying the way the crop is grown, for example by increasing vegetational diversity near or in the crop (Altieri and Letourneau, 1984; van Emden, 1990), or by changing some cropping practice, such as sowing date or cultivation. Frequently, the most important change is to modify pesticide use and, in particular, to reduce it to a level determined by monitoring pest numbers. Pesticide application can be particularly damaging to natural enemies (Waage, 1989).

All other forms of biological control incur the cost of producing and releasing natural enemies. Hence, conservation of natural enemies should be considered first.

Augmentation and Inundative Release

Where conservation is insufficient, natural enemies may need to be added to the crop system. This might involve the augmentation of field populations of natural enemies with laboratory-reared individuals released early in a season, to check the pest population before it reaches damaging levels. Alternatively, it may involve massive and regular applications of natural enemies as biopesticides through a crop season. The method chosen depends very much on the ecology of the natural enemy. For instance, insect pathogens, with their relatively poor capacity for transmission, are often best used inundatively as biopesticides.

The augmentation of natural enemies can be expensive, and therefore will often find itself in competition with the use of chemical pesticides. Its advantages relative to chemicals are those identified generally for biological control above. Its disadvantages are the short shelf-life of living organisms and their variable effect in the field, due to climate and substrate. For this reason, we usually find augmentation adopted by organized farming systems, such as plantations and glasshouses, where there is a continuous and dependable demand for natural enemies and a well-organized means of distributing them. Further, many plantations crops, like sugar-cane and citrus, are less subject to climatic variation and human perturbation than field crops.

Classical Biological Control

The third approach to biological control is relevant to the special situation where the pest in question is exotic to the region. Exotic pests frequently invade regions without their adapted, natural enemy complex. It is often effective to introduce and establish these specific natural enemies of the pest from its area of origin. This method is frequently called classical biological control, because it has been employed for over a century. Classical biological control can also be directed against native pests, where exotic natural enemies can be found which attack them, although this is less common.

The aim of classical biological control is to reduce population densities of a pest, through the importation and establishment of natural enemies, to a level at which the pest no longer causes economic damage. The implementation of classical biological control involves six steps (Fig. 16.1).

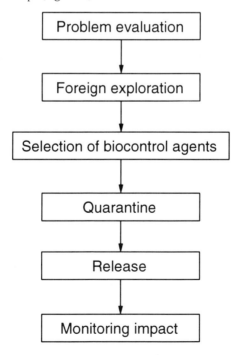

Fig. 16.1. Six steps in the implementation of classical biological control.

It is important first to evaluate the pest problem, to establish the identity of the pest and its probable area of origin. In addition, information must be gained on the pest's ecology in its exotic range and the natural enemies that may be associated with it there. Then, foreign exploration may be made for these natural enemies, involving quantitative surveys to assess the natural enemy complex and its impact.

Promising natural enemies are then selected from this complex for introduction as classical biological control agents. Selection requires making educated guesses about which natural enemies will be most effective. Predicting effectiveness can involve experimentation and theory and, not surprisingly, the interpretation of predictive methods makes this step the most controversial in the process.

Quarantine is the process that natural enemies must go through once they have been selected. For insect natural enemies, this means removing any hyperparasitoids, plant pathogens and insect pathogens from the culture. Following quarantine, the candidate natural enemies are released, and both natural enemy and pest populations are monitored, to document the establishment of the natural enemy and to evaluate its impact on the pest population.

DECISION MAKING IN THE CONSERVATION OF NATURAL ENEMIES

The different approaches to biological control involve different challenges in decision making, and the first decision concerns the choice of approach. Since both other approaches involve the costs of producing and releasing natural enemies, conservation of existing agents is a natural first choice. A conservation approach may also be dictated by the nature of the pest's environment and the farming system. Not only can this be unsuitable for augmentative control but it may also preclude classical biocontrol introductions. For example many attempts at classical biological control of the olive moth *Prays oleae* have failed, probably because of the very hot summers and continued ploughing of the soil in Mediterranean olive groves. Consequently future IPM strategies are focusing on conservation of natural enemies capable of surviving in this demanding environment (Kidd and Jervis, 1993).

In conservation, research often leads to a method which can be implemented without the need for regular decisions by the farmer or extension worker. Thus, modification of a cropping practice may be a once-and-for-all change. However, regular decision making is required when we seek to protect natural enemies by reducing pesticide application. The interaction between pesticides, pests and natural enemies is highly complex, as illustrated in Fig. 16.2.

Both pesticides and natural enemies reduce pest populations (arrows in Fig. 16.2). By applying pesticides, the farmer reduces the pest population, but also may reduce natural enemy populations, releasing pests from their controlling effect. Pesticides may also reduce natural enemies indirectly as a result of the enemies' numerical response to lowered pest densities. Depending upon the relative strength of all four vectors, pesticide application could enhance biological control, have no effect on it, effectively replace it, or lead to the disaster of resurgence (Waage, 1989).

With appropriate information on the vector arrows illustrated in Fig. 16.2, farmers can, in principle, make decisions about whether to use pesticides, when to use them and which pesticides to use. However, this information is difficult to come by. One

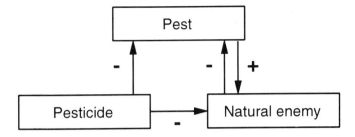

Fig. 16.2. Interactions between pesticides, pests and natural enemies.

way of approaching this problem is described in Chapter 8. A simulation model of the rice brown planthopper (BPH) has been used to investigate the disruptive effect of pesticides, in causing a reduction in the mortality of BPH caused by natural enemies. It is used to examine the effect of the type and time of pesticide spray, as well as natural enemy recovery time, on BPH resurgence. As a result of this work, the resurgence risks of a particular type and time of spray can be assessed.

An even simpler simulation model has been used to predict the effect of chemical or cultural control measures on the natural control of grass grub *Costelytra zealandica* in New Zealand by protozoan diseases (Barlow and Jackson, 1992). The model uses a standard difference equation for density-dependent changes in the grub population (*N*) from year to year, together with an equation relating the proportion of grubs diseased in one year (*D*) asymptotically to the amount of potential inoculum in the previous year, which is proportional to the density of diseased grubs in that year (*N* × *D*):

Grubs: $$N_{t+1} = aN_t^{1-b}(1 - D_t)^c$$
Proportion diseased: $$D_{t+1} = fN_tD_t/(1 + fN_tD_t)$$

where *a*, *b*, *c* and *f* are constants and *t* is the generation and year. *c* depends on the order of density-dependence and mortality from the disease, and on the number of times the disease cycles per generation of the grubs (Barlow and Jackson, 1992).

The model was programmed in Turbo Pascal for interactive use, with simple and visually appealing colour graphics (Fig. 16.3). This way of presenting biological simulation models greatly helps in their adoption as decision support tools, because it allows decision makers to understand the biology, experiment for themselves and see quickly the results of any action they decide upon.

In this case the model dramatically illustrates the consequences of disrupting natural biological control by the application of a chemical spray, namely a pest outbreak three years later (Fig. 16.3). The outbreak occurs not because the pesticide directly affects the natural enemy, in this case the protozoan diseases, but because of the reduction in the pest density which leads in turn to a depression in disease levels. This pathway is shown by the arrows in Fig. 16.2, and the reduction of disease levels after the spray can be seen in the model (dotted line in Fig. 16.3). In cases like this any control, whether chemical or cultural, will disrupt natural biological control. The model further suggests that the cost of damage to pasture by grass grub (shown in the left-hand box in Fig. 16.3), over the 20-year period, is the same as if no pesticide had been applied, and in the latter case there is the additional cost of the spray (shown in the second box of Fig. 16.3).

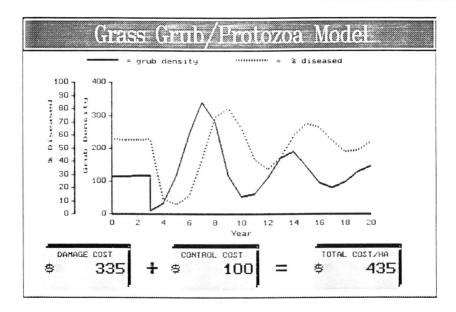

Fig. 16.3. Representative output from a decision-support simulation model for the effects of chemical and cultural controls on populations of the New Zealand grass grub naturally regulated by protozoan diseases. The simulation shows the results of applying chemical control causing 90% grub mortality in year 3, to a population initially at equilibrium. The boxes show the costs of grub damage to pastures accumulated over 20 years.

The usefulness of such models as decision tools lies less in their value as tactical aids than in helping growers or managers gain an awareness of ecological function and of the likely general consequences of their actions. In practice the main problem with grass grub is that natural biological control is imperfect, particularly under the disturbing influence of weather (dry summers) and farm management practices. Ploughing a field, for example, creates a situation with low grub densities and low disease levels. This leads to an outbreak in three years or four years after ploughing. However, the above model shows that if the grower attempts to avert this by applying control to the increasing population, the result is only to delay but not prevent the outbreak. Repeated control is necessary to eliminate it altogether, but this gradually reduces disease levels and therefore sensitizes the system to any failure in control; when it ceases to be applied the outbreak is much more severe than if no additional control had been used. The graphical model makes outcomes such as this very obvious to the manager.

The inclusion of natural enemies into mathematical models for action thresholds usually involves the integration of models for pest population growth with those for natural enemy dynamics (Stimac and O'Neil, 1985). While modelling of this kind gives insights on a process which might be prohibitive to explore empirically, the final decision tools derived from it must be accessible to the user.

Thus, in the case of BPH in Malaysia, a monitoring system has been developed in which farmers count both pests and natural enemies, and determine whether to spray on the basis of the ratio of their abundance. This presumes that farmers can distinguish pests and natural enemies, an important first step requiring training.

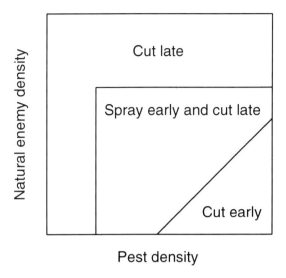

Fig. 16.4. Generalized multi-dimensional threshold decision graph sowing different strategies for different combinations of pest and natural enemy densities.

A somewhat more complex method has been used to determine the best combination of biological, chemical and cultural methods for alfalfa weevil control, expressing these as multi-dimensional economic thresholds (Shoemaker, 1979). Based on dynamic programming, which includes the temperature-dependent population dynamics of the weevil and its parasite, the final decision aid is a simple graph showing, for combinations of weevil and parasite densities (represented by the horizontal and vertical axes), the optimum date for harvesting the alfalfa and whether insecticide use is justified (areas on the graph). A generalized form of this kind of graph is shown in Fig. 16.4. There are three such graphs, for warm, average and cool seasons, and insecticide is most likely to be recommended in warm seasons and when weevil densities are high and parasite densities low.

A similar decision-making graph is described in Asquith *et al.* (1980) for biological and chemical control of European red mite in Michigan apple orchards. Again a graph of predatory mite numbers (*Amblyseius fallacis*) against spider mite numbers per leaf is divided into regions, each of which denotes a probability of natural biological control occurring and a management recommendation. These recommendations include no action, if predator numbers are high relative to those of the prey, waiting and monitoring again, or spraying at different rates with an acaricide to which the predator is resistant.

DECISION MAKING IN AUGMENTATION

The repeated release of natural enemies, like the use of chemical pesticides, is best done on an at-need basis which ensures minimal use to get the necessary control of the pest, and hence minimal cost to the farmer. But natural enemies have some problems and some striking advantages which makes them different from chemicals and more of a challenge to the creation of decision tools.

Many natural enemies act slowly relative to chemicals, and this delay must be anticipated in decision making about their use. On the positive side, they are living organisms which reproduce, and the impact of a released population may be compounded over successive natural enemy generations. To make most use of this remarkable property of biological control agents, we would need to predict pest and natural enemy population trends from some initial densities, a process similar to that discussed in the previous section, and to derive simple decision tools from this.

Needless to say, such an exercise is complex and has to date attracted little attention. By and large most augmentation of natural enemies, like most chemical pesticide use, still follows a strategy of 'calendar' or 'insurance' application of very large numbers of mass-reared predators, parasitoids or pathogens, which ignores the agents' own reproductive capacity.

Recent trial and error studies have shown the value of building population dynamic predictions into augmentation decisions. The moth egg parasitoids, *Trichogramma* spp., the most widely augmented insect natural enemy, are a prime example. In countries such as China, stem borers in sugar-cane are controlled by calendar releases of 250-500,000 *Trichogramma* per hectare every 2–3 weeks during the season of infestation. Studies have shown, however, that the same degree of control can be achieved by releasing much smaller numbers at a strategic point earlier in the season. In one case (Guo, 1986), two releases of 75,000 wasps each early in March gave the same degree of control as 8–12 releases of over 200,000 at intervals from April to July.

This method works because the absolute rate of increase of an initial, low pest population is very much lower than that of a later, higher population. Fewer natural enemies are therefore needed to prevent growth of an early population. The critical challenge for decision making, therefore, is to time releases early enough to ensure that the pest population is small enough to be controlled (balancing the ratio of pest and natural enemy is crucial here), but late enough to ensure that there are enough pests around for easy establishment and build-up of natural enemies.

The value of mathematical population modelling to the development of decision rules for augmentation has been appreciated for some time. For example, Yano (1989) used a simulation model of the greenhouse whitefly and the parasite *Encarsia formosa* to show that a given number of parasites was best applied over a number of separate occasions spanning a long initial period. Smith and You (1990) found, also by simulation modelling, that the best timing for inundative release of the egg parasite *Trichogramma minutum* against spruce budworm was 14 days after appearance of the first budworm egg masses, and that either two releases or a single release with staggered emergence of the parasite were required to control the pest.

This approach can be applied to augmented pathogens as well as insects. For example, the New Zealand grass grub mentioned above can be controlled by addition of a commercially produced bacterium *Serratia entomophila*, which occurs naturally in part of the pest's range. Decision models based on simulation, and very similar to the one described above for protozoan diseases, are being used to assess the costs and benefits of augmentation, and especially the circumstances under which augmentation gives most economic benefit (Barlow and Jackson, 1992). Such circumstances include weather, age of the field, history of previous damage, or grub densities which can be measured in the field.

DECISION MAKING IN CLASSICAL BIOLOGICAL CONTROL

The Decision Problem

The challenge of decision making in classical biological control is quite different from that discussed above. With conservation and augmentation, we can use decision tools to tell us the impact of a certain number of indigenous natural enemies on a particular pest population at some point in the future. With classical biological control, estimating impact is also the aim, but we have the disadvantage that the species whose impact we need to estimate have never existed in the crop in question: they are exotic species destined for introduction. The decision we must make is which species to introduce. Without knowing how they perform where we want to release them, this becomes a difficult and speculative task. The rest of this chapter will focus on this challenge.

Exploratory programmes for biological control agents frequently turn up between 10 and 20 species of potential natural enemies for introduction against an exotic pest. Not all of these are suitable candidates for biological control. Some will be too generalized in their host range to be considered safe in a new environment. Safety testing is essential and time consuming – in the case of agents for weed control, it may take 2–4 years for a particular target weed and its natural enemies.

Having screened out natural enemies unlikely to be safe we may still be left with a reasonable number of candidate agents. Classical biological control programmes are usually very expensive, given the high costs of working in different countries and safety testing of agents. As a result, the funding for them is usually not sufficient to process all potential and safe agents, nor would the recipient country have the patience to wait long for a trial and error solution to an urgent exotic pest problem.

Instead, we need to select from our safe agents, the ones most likely to succeed. The logical aim here is to have a ranking of agents according to their potential value, such that a programme can work from the best to the worst, as long as funding holds out. Concentrating on an inferior agent not only wastes time and money, it may also prevent future redress: recent studies of biological control programmes have suggested that, with the introduction of each agent species, the addition of another species gets more difficult (Ehler and Hall, 1982). This may reflect problems in establishing exotic agents on pest populations which are smaller, because they are already being partially controlled (for instance, by the first agents introduced). Thus, there may only be one, or a few, chances to get it right.

Several factors are usually considered in ranking agents:

- Climatic and phenological adaptation of the agent to the exotic distribution of the pest.
- Precedent for previous success with that group of agents, from the history of biological control.
- Ecological properties of the agent or the pest likely to indicate success.

The selection of classical biological control agents has all the properties of a system which could benefit from decision tools. A number of criteria for selection must be applied, leading to a decision on the ranking of potential agents. We have portrayed these steps as a sequence, but in fact, the procedure can be far more complex. For

instance, it is not worthwhile to expend several years in safety testing a weed control
agent if it is not promising on other grounds. We may wish, therefore, to apply our ecological selection criteria first, before safety testing.

Let us now look at the current and potential role that different decision tools can
play.

Climatic Models and Databases

Information used in agent selection can be put into a database or model format to facilitate decision making. Thus, practitioners often try to select sites for exploration by
matching the climate of the intended area of introduction of the natural enemy with a
part of its area of origin. In temperate areas in particular, matching the climate and phenological attributes of the agent to the exotic distribution of the pest is most important
because periods of activity and diapause may differ between sites of origin and introduction, but tropical systems also can have extreme seasonality where phenological adaptation may be crucial.

In the past, practitioners have used atlases of climatic data to make these comparisons. Today, this information can be computerized and its value enhanced to facilitate
exploratory programmes. The climatic matching programme, CLIMEX, described in
Chapter 6, is such an initiative, and it has been of value in identifying sites for exploration for exotic insect and weed pests (Worner et al., 1989). Another climatic matching
database has been used to select areas of exploration in Latin America for predatory
mites to introduce into Africa against the cassava green mite, *Mononychellus tanajoa*
(Yaninek and Bellotti, 1987).

The use of precedents is particularly important to the selection of classical biological
control agents. There is a long record of classical biological control effort which, while
often poorly documented, gives the clear indication that certain pests and natural enemies are associated with more success than others.

To help better utilize this information in decision making, a number of reviews have
been made of classical biological control around the world, usually with a regional emphasis (Clausen, 1978; Cock, 1985; Cameron et al., 1989). More recently, IIBC has developed
a database of introductions of arthropod natural enemies for control of invertebrate pests,
called BIOCAT. This database is easily searched on a number of criteria (for example
target pest, country, natural enemy, data, level of success) and cross referenced to original literature. BIOCAT helps IIBC to accelerate evaluation of potential agents for classical
biological control, and to respond quickly to queries about the potential for classical biological control of new pest problems.

Analysis of BIOCAT has provided useful, general information. It has shown, for
instance, that about 40% of all established biocontrol agents contribute substantially to
control of the pest (Waage and Greathead, 1989). Of the pest groups involved in
these programmes, the Homoptera have been by far the most targeted and most successfully controlled taxon (Greathead, 1989).

A number of academic groups have also established databases of classical biological
control, and some of these have been applied to explore more fundamental ecological
questions. Thus, Southwood (1977) has argued that natural enemies are likely to be
most effective in habitats or crops of intermediate stability, i.e. between the extremes

of short-cycle, annual field crops and permanent forests. The records from databases tend to support this, with greater success in orchard crop systems than in seasonal field crops (Hall and Ehler, 1979; Hall *et al.*, 1980; Greathead and Waage, 1983). Databases have also been used to explore the relative value of introducing exotic agents to control exotic vs. indigenous pests (Hokkanen and Pimentel, 1984).

Ecological Models

Perhaps the most problematic, and potentially most valuable, decision tool for selecting agents is the application of population ecology. Over the past 50 years, ecologists have made increasing efforts to apply theory, and particularly mathematical population modelling, to the selection of biological control agents. This has not been an entirely altruistic exercise. Classical biological control programmes, because they often involve a single pest species and one or a few natural enemies in isolation in a new region, resemble the simple predator–prey systems that ecologists have worked with in the laboratory and in modelling. Hence, they are excellent natural experimental systems for the evaluation of ecological theory.

Waage (1990) has suggested that it is possible to classify two approaches by which ecologists have contributed to the selection of biological control agents: the **reductionist approach** and the **holistic approach**.

The reductionist approach develops criteria for selection based on the idea that a natural enemy can be described as a set of component characters which influence its impact. The development of simple, analytical models for predator–prey interactions (as described in Chapter 7) lies at the heart of the reductionist approach to selecting insect biocontrol agents. Reductionist criteria are derived from the parameters of these models, particularly those which appear important to pest depression and regulation. Searching efficiency (or area of discovery), handling time, aggregation and mutual interference have all been suggested as criteria for selecting agents. An example of how such criteria might be used would be to execute a functional response experiment in order to calculate and compare the searching efficiency of three candidate natural enemies, and to select for introduction the agent with the highest value.

In practice, reductionist criteria have rarely been used in selecting agents. Problems with the reductionist approach are several (Waage, 1990; Waage and Mills, 1992), and include the considerable difficulty in measuring key parameters in laboratory and field and the presumed independence of different selection parameters from each other, and from other processes influencing the pest's population ecology.

A distilled version of this approach, usually represented by lists prefaced by 'Good natural enemies should have the following properties . . . ', is to be found today in most textbook treatments of insect biological control. As a summary of the results of experience and simple predator–prey theory, it has broad, educational value.

A somewhat similar but more empirical approach has been taken over the years to selection of weed control agents. Here, we often find that a number of candidate agents exist with very different biologies, in terms of the plant parts they attack and the kind of damage they do, leading to some interesting challenges for selection. This approach has also relied on key life-history parameters of the agent (e.g. fecundity, site of feeding, feeding habits), but it has also incorporated more ecological criteria (for example phenology), into a scoring system for comparing a suite of potential agents (Harris, 1973; Goeden, 1983). This scoring system has had practical use, and useful refinement, over the years, with some success.

The term 'holistic' has been applied to different kinds of methods for agent selection which look more at interactions than specific properties, and consider how a particular natural enemy might fit into the life history of the pest in its exotic range, with all the other factors which affect this. In particular, these methods have looked at the interactions of mortalities caused by the natural enemy and other factors, including other introduced natural enemies.

Recent experience in biological control has highlighted the importance of these interactions. For instance, Hill (1988) suggests that the impact of the introduced larval parasitoid, *Apanteles rufricrus*, on populations of the moth, *Mythimna separata*, may have been reduced by density-dependent parasitism by a later-acting, indigenous pupal parasitoid. Similarly, the contribution of egg parasitoids may be reduced by density-dependent mortalities acting on young larval stages of their moth hosts (van Hamburg and Hassell, 1984). In principle, such over-compensating density-dependent mortalities can cause natural enemies which precede them to actually increase pest abundance (May and Hassell, 1981).

A holistic approach is also relevant to the decision of how best to combine natural enemies where more than one species is introduced. It can be applied to determine which combinations of species, acting in different ways, will be most effective in reducing pest numbers, and which combinations may actually lead to interference between agents.

From these two traditional perspectives, reductionist and holistic, is emerging a consensus that we need to know about both the properties of agents and their interactions with each other and other factors in order to most accurately predict their fate after introduction. This means we must consider a number of species, a number of stages in the life history of each species, the mortalities acting at these stages and how this all integrates over time. Needless to say, such a challenge lends itself to a mathematical modelling approach, and one which considers not only reductionist criteria but age-specific phenomena and the interactions of mortalities. At the same time, some simplicity in approach is required if these models are to be useful and robust.

Over recent years, there have been a number of attempts to model biological control in this way. These have employed a range of modelling formats, from analytical to simulation, and include studies of the winter moth (Hassell, 1980), red scale (Murdoch *et al.*, 1987), cassava mealybug (Gutierrez *et al.*, 1988), mango mealybug (Godfray and Waage, 1991), and *Sitona* weevil (Barlow and Goldson, 1993). While most of these models have been retrospective, they reveal the potential for using this kind of model in a prospective manner.

For example, a model is currently being developed to predict the impact of a newly released parasitoid, *Microctonus hyperodae*, on Argentine stem weevil in New Zealand. This is possibly the first time that an attempt has been made to predict the outcome of a classical biological control introduction in advance. However, the main aim in this case is to help understand the ecological basis for eventual success or failure of the agent in general and of the seven distinct ecotypes introduced. The model was not used prospectively in agent selection. A recent programme against mango mealybug was one of the first to use prospective modelling as an exercise in agent selection, and this study will be described in more detail.

Case Study – Biological control of *Rastrococcus invadens*

By giving a specific case study in classical biological control, we will try to show in a practical way how decisions are made and how predictive modelling can guide decisions in agent selection.

The mango mealybug, *Rastrococcus invadens*, was first discovered in Ghana and Togo around 1981 (Agounké *et al.*, 1988) causing damage to fruit trees, particularly mango and citrus. It feeds by sucking sap from the undersurface of the leaves and its honeydew production leads to the growth of sooty moulds on the foliage. The net result is a marked reduction in fruit production, a resource providing an essential source of both energy and vitamins in the diet of the local population. Apart from being a staple food, it has been estimated that the annual value of the mango crop in Togo is £1.95 million (J.M. Voegele, unpublished data). The extent of economic losses are difficult to assess, given the traditional nature of horticultural systems in the region, and remain largely unquantified although losses of up to 80% from mangoes were noted in Ghana (Willink and Moore, 1988). Thus potential annual losses exceeding £1.5 million for individual nations indicate that the problem was sufficiently severe to warrant a classical biological control programme. Analysis of the record of classical biological control indicated that mealybugs are very promising targets.

The mealybug was first confused with another, Asian species, but subsequently proved to be new to science. It was described in 1986 (Williams, 1986) and a reassignment of museum material revealed that its probable area of origin was South and Southeast Asia. Accordingly, IIBC initiated surveys for natural enemies in India and Malaysia in July 1986. From a series of three *Rastrococcus* species found feeding on mangoes in India, about 15 species of parasitoids and predators were recorded (Narasimham and Chacko, 1988), most unknown to science. These included several encyrtid parasitoids, which have proved to be the most successful agents used against mealybugs in classical biological control (Moore, 1988). Host specificity of some encyrtid parasitoids of other *Rastrococcus* species revealed that they did not attack *R. invadens*. Therefore, further work concentrated on the parasitoids of *R. invadens*.

These initial surveys indicated that *R. invadens* has two encyrtid parasitoids, *Gyranusoidea tebygi* and *Anagyrus mangicola*. Field observations in India, where *R. invadens* was not abundant at any localities, suggested that parasitism seldom exceeded 20% and that *A. mangicola* was the dominant parasitoid (Narasimham and Chacko, unpublished). Further studies of the biology of *R. invadens* and its parasitoids at IIBC in the United Kingdom (Willink and Moore, 1988; Moore, unpublished) showed that both *G. tebygi* and *A. mangicola* were promising agents. *G. tebygi* proved to be more easily reared than *A. mangicola* and so was selected as the first candidate for introduction.

The first releases of *G. tebygi* were made near Lomé in Togo in November 1987. The parasitoid readily established from releases of as few as 20–30 individuals in some sites and after only one year the parasitoids had migrated up to 100 km from the original release sites (Agricola *et al.*, 1989). Releases of *G. tebygi* were later extended to Benin, Ghana, Nigeria, Zaire and Gabon, and in all cases the parasitoids have established. The current indications are that good control of the mealybug has been achieved throughout the areas colonized and monitoring of the impact of the releases is in progress.

In 1987, work began on a population model to predict the potential impact of different natural enemies of *R. invadens*. The model was constructed for *G. tebygi* and *A.*

mangicola, using basic data collected in India and the United Kingdom, and some information on survivorship of the mealybug in the field in Togo. The emphasis in modelling was on the incorporation of important but easily measurable life-history parameters, which could be realistically collected in a few months of field and laboratory study. Key parameters included the stage of host attacked by the different parasitoids, age-specific development rates for hosts and parasitoids, age-specific survivorship of hosts in the field, and adult longevities and daily oviposition rates. Difficult criteria, such as searching efficiency, were treated as variables.

Due to the urgency of the problem, *G. tebygi* was released before the model was complete. However, once the model was complete, it predicted that *G. tebygi* would cause greater depression of mealybug populations than *A. mangicola*, and that the latter would not contribute to substantial additional depression when both were used (Godfray and Waage, 1991). Subsequent laboratory population studies in the United Kingdom supported these findings: *A. mangicola* was less effective at suppressing colonies of *R. invadens* over ten generations and went extinct in the presence of the superior *G. tebygi* (Moore and Cross, unpublished results). As a result of these studies, any new introductions in affected countries will focus on *G. tebygi*, but opportunities will be sought to test *A. mangicola* in combination with it.

Summary

In this chapter we have considered how models, or model-based graphs, charts and recommendations, can help in making both tactical and strategic decisions involving biological control.

Examples of tactical aids are the multi-dimensional economic thresholds, which suggest whether existing natural enemies need to be supplemented with chemical or cultural controls in particular circumstances, having regard to the effects of such controls on the natural enemies.

Aids to strategic decision making include the brown planthopper and grass grub simulation models, which show the consequences of failing to conserve natural enemies and the economic benefits of pathogen augmentation, models suggesting the optimal timing for inundative releases of *Trichogramma* and *Encarsia*, and prospective models for the selection of classical biological control agents.

In most of these cases it is the result of the model, often expressed as a graph, table or a simple recommendation, which comprises the actual decision tool. However, the grass grub model shows the potential for simulation models also to be used as interactive decision tools in themselves, allowing the manager to experiment and to pose and obtain answers to an effectively unlimited array of 'what if?' questions.

Perhaps the newest and most exciting area for decision making in biological control lies in the selection of classical biocontrol agents. However, a major challenge in developing useful prospective models for the selection of agents is the combination of ecological complexity and the limited time frequently available for decision making. This dictates a pragmatic approach, using fairly holistic models of intermediate complexity (Godfray and Waage, 1991).

Alternatively, the models' aims may be made more modest, constrained to particular aspects of the selection decision such as the way in which candidate agents are likely to perform in the climatic environment of the target pest. Here phenological models are capable of predicting likely generation numbers for the natural enemy and degree of synchrony with the pest, and they can be constructed relatively quickly since the data they require on the natural enemy can be obtained in the laboratory or under quarantine. However, in the future better decision tools for classical biological control, whether in the form of models or general rules, are likely to require a sounder knowledge base and a correspondingly greater emphasis on the ecological analysis and evaluation of practical biological control programmes, in which modelling plays a key role (Barlow and Goldson, 1990, 1993).

References

Agounké, D., Agricola, U. and Bokonon-Ganta, H.A. (1988) *Rastrococcus invadens* Williams (Hemiptera: Pseudococcidae), a serious exotic pest of fruit trees and other plants in West Africa. *Bulletin of Entomological Research* 78, 695–702.

Agricola, U., Agounké, D., Fischer, H.U. and Moore, D. (1989) The control of *Rastrococcus invadens* (Hemiptera: Pseudococcidae) in Togo by the introduction of *Gyranusoidea tebygi* Noyes. *Bulletin of Entomological Research* 79, 671–678.

Altieri, M.A. and Letourneau, D.K. (1984) Vegetation diversity and insect pest outbreaks. *CRC Critical Reviews in Plant Sciences* 2, 131–169.

Asquith, D., Croft B.A., Hoyt, S.C., Glass, E.H. and Rice, R.E. (1980) The systems approach and general accomplishments toward better insect control in pome and stone fruits. In: Huffaker. C.B. (ed.) *New Technology of Pest Control*. Wiley Interscience, New York. pp. 249–318.

Barlow, N.D. (1993) Applications and challenges in pest population modelling. In: Leather, S., Walters, R., Mills, N. and Watt, A. (eds) *Individuals, Populations and Patterns*. Intercept, Andover, UK.

Barlow, N.D. and Goldson, S.L. (1990) Modelling the impact of biological control agents. *Proceedings of the 43rd New Zealand Weed and Pest Control Conference*, pp. 282–283.

Barlow, N.D. and Goldson, S.L. (1993) A modelling analysis of the successful biological control of *Sitona discoideus* (Coleoptera: Curculionidae) by *Microctonus aethiopoides* (Hymenoptera: Braconidae) in New Zealand, *Journal of Applied Ecology* 30 (in press).

Barlow, N.D. and Jackson T.A. (1992) Modelling the impact of diseases on grass grub populations. In: Jackson, T.A. and Glare, T.R. (eds) *Use of Pathogens in Scarab Pest Management*. Intercept, Andover, UK, pp. 127–140.

Cameron, P.J., Hill, R.L., Bain, J. and Thomas, W.P. (eds) (1989) *A Review of Biological Control of Invertebrate Pests and Weeds in New Zealand 1874 to 1987*. CAB International and DSIR, Wallingford, UK.

Clausen, C.P. (ed.) (1978) *Introduced Parasites and Predators of Arthropod Pests and Weeds: A World Review*. Agricultural Handbook No. 480, Agriculture Research Service, United States Dept of Agriculture.

Cock, M.J.W. (ed.) (1985) *A Review of Biological Control of Pests in the Commonwealth Caribbean and Bermuda up to 1982*. Technical Communication No. 9, CAB International, Wallingord, UK.

Ehler, L.E. and Hall, R.W. (1982) Evidence for competitive exclusion of introduced natural enemies in biological control. *Environmental Entomology* 11, 1–4.

Godfray, H.C.J. and Waage, J.K. (1991) Predictive modelling in biological control: the mango mealy bug (*Rastrococcus invadens*) and its parasitoids. *Journal of Applied Ecology* 28, 434–453.

Goeden, R.D. (1983) Critique and revision of Harris' scoring system for selection of insect agents in biological control of weeds. *Prot. Ecology.* 5, 287–301.

Greathead, D.J. (1989) Biological control as an introduction phenomenon: a preliminary examination of programmes against the Homoptera. *The Entomologist* 108, 28–37.

Greathead, D.J. and Waage, J.K. (1983) *Opportunities for Biological Control of Agricultural Pests in Developing Countries*. World Bank Technical Paper No. 11, The World Bank, Washington, DC.

Guo, Ming-Fang (1986) New method of *Trichogramma* utilization. In: *Trichogramma and Other Egg Parasites*. IInd International Symposium, Guangzhou (China) Ed. INRA, Paris.

Gutierrez, A.P., Neuenschwander, P., Schultess, F., Herren, H.R., Baumgaertner, J.V., Wermelinger, B., Lohr, B. and Ellis, C.K. (1988) Analysis of biological control of cassava pests in Africa. II. Cassava mealybug, *Phenacoccus manihoti*. *Journal of Applied Ecology* 25, 921–929.

Hall, R.W. and Ehler, L.E. (1979) Rate of establishment of natural enemies in classical biological control. *Bulletin of the Entomological Society of America* 25, 280–282

Hall, R.W., Ehler, L.E. and Bisabri-Ershadi, B. (1980) Rate of success in classical biological control of arthropods. *Bulletin of the Entomological Society of America* 26, 111–114

Harris, P. (1973) The selection of effective agents for the biological control of weeds. *The Canadian Entomologist* 105, 1495–1503.

Hassell, M.P. (1980) Foraging strategies, population models and biological control: a case study. *Journal of Animal Ecology* 49, 603–628.

Hill, M.G. (1988) Analysis of the biological control of Mythimna separata (Lepidoptera: Noctuidae) in New Zealand. *Journal of Applied Ecology* 25, 197–208.

Hokkanen, H. and Pimentel, D. (1984) New approach for selecting biological control agents. *The Canadian Entomologist* 116, 1109–1121.

Kidd, N.A.C. and Jervis, M.A. (1993) Development of management strategies for the olive moth – the role of modelling. In: Leather, S., Walter, R., Mills, N. and Watt, A. (eds) *Individuals, Populations and Patterns*. Intercept, Andover, UK.

May, R.M. and Hassell, M.P. (1981) The dynamics of multiparasitoid–host interactions. *The American Naturalist* 117, 234–261.

Moore, D. (1988) Agents used for biological control of mealybugs (Pseudococcidae) *Biocontrol News and Information* 9, 209–225.

Murdoch, W.W., Nisbet, R.M., Blythe, S.P., Gurney, W.S. and Reeve, J.D. (1987) An invulnerable age class and stability in delay-differential parasitoid–host models. *The American Naturalist* 129, 263–282.

Narasimham, A.U. and Chacko, M.J. (1988) *Rastrococcus* spp. (Hemiptera: Pseudococcidae) and their natural enemies in India as potential biocontrol agents for *R. invadens* Williams. *Bulletin of Entomological Research* 78, 703–708.

Shoemaker, C. (1979) Optimal management of an alfalfa ecosystem. In: Holling, C.S. and Norton, G.A. (eds) *Pest Management*. IIASA, Laxenburg, Austria., pp. 301–315.

Smith, S.M. and You, M. (1990) A life system simulation model for improving inundative releases of the egg parasite *Trichogramma minutum* against the spruce budworm. *Ecological Modelling* 51, 123–142.

Southwood, T.R.E. (1977) Habitat, the templet of ecological strategies? *Journal of Animal Ecology* 46, 337–365.

Stimac, J.L. and O'Neil, R.J. (1985) Integrating influences of natural enemies into models of crop/pest systems. In: Hoy, M.A. and Herzog, D.C. (eds) *Biological Control in Agricultural IPM Systems*. Academic Press, New York, pp. 323–346.

van Emden, H.F. (1990) Plant diversity and natural enemy efficiency in agroecosystems. In: Mackauer, M., Ehler, L.E. and Roland, J. (eds) *Critical Issues in Biological Control*. Intercept Ltd, Andover, UK, pp. 63–80.

van Hamburg, H. and Hassell, M.P. (1984) Density dependence and the augmentative release of egg parasitoids against graminaceous stalkborers. *Ecological Entomology* 9, 101–108.

Waage, J.K. (1989) The population ecology of pest–pesticide–natural enemy interactions. In: Jepson, P. (ed.) *Pesticides and Non-target Invertebrates*. Intercept Ltd, Andover, UK, pp. 81–93.

Waage, J.K. (1990) Ecological theory and the selection of biological control agents. In: Mackauer, M., Ehler, L.E. and Roland, J. (eds) *Critical Issues in Biological Control*. Intercept Ltd, Andover, UK, pp. 135–158.

Waage, J.K. and Greathead, D.J. (1989) Biological control: challenges and opportunities. In: Wood, R.K.S. and Way, M.J. (eds) *Biological Control of Pests, Pathogens and Weeds: Developments and Prospects*. University Press, Cambridge, pp. 1–18.

Waage, J.K. and Mills, N.J. (1992) Biological control In: Crawley, M.J. (ed.) *Natural Enemies, The Population Biology of Predators, Parasites and Diseases*. Blackwell Scientific Publications, Oxford.

Williams, D.J. (1986) *Rastrococcus invadens* sp. n. (Hemiptera: Pseudococcidae) introduced from the

Oriental Region to West Africa and causing damage to mango, citrus and other trees. *Bulletin of Entomological Research* 76, 695–699.

Willink, E. and Moore, D. (1988) Aspects of the biology of *Rastrococcus invadens* Williams (Hemiptera: Pseudococcidae), a pest of fruit crops in West Africa, and one of its primary parasitoids, *Gyranusoidea tebygi* (Hymenoptera: Encyrtidae). *Bulletin of Entomological Research* 78, 709–715.

Worner, S.P., Goldson, S.L. and Frampton, E.R. (1989) Comparative ecoclimatic assessment of *Anephes diana* (Hymenoptera: Mymaridae) and its intended host, *Sitona discoides* (Coleoptera: Curculionidae), in New Zealand. *Journal of Economic Entomology* 82, 1085–1090.

Yaninek, J.S. and Bellotti, A.C. (1987) Exploration for natural enemies of cassava green mites based on agrometeorological criteria. In: Rijks, D. and Mathys, G. (eds) *Proceedings of the Seminar on Agrometeorology and Crop Protection in the Lowland Humid and Subhumid Tropics, Cotonou, Benin.* World Meteorological Organization, Geneva.

Yano, E. (1989) A simulation study of population interaction between the greenhouse whitefly, *Trialeurodes vaporariorum* (Homoptera: Aleyrodidae) and the parasitoid *Encarsia formosa* (Hymenoptera: Aphelinidae) II. Simulation analysis of population dynamics and strategy for biological control. *Research in Population Ecology* 31, 89–104.

17 Extension Techniques for Pest Management

C. Garforth

*Agricultural Extension and Rural Development Department, No. 3 Earley Gate,
University of Reading, Whiteknights Road, Reading RG6 2AL, UK*

Introduction

Most of the tools described in this book are designed for use by researchers, strategic decision makers or people whose job it is to advise farmers. However, most pest management action is taken by individual farmers. They make decisions in response to the situation as they perceive it in their crops and animals, and on the basis of the knowledge and information they have. If the researcher's, planner's and adviser's decision tools are to have any influence on farm level practices, a bridge is needed: national strategies have to be translated into local action; advisers must communicate their advice in such a way that farmers can understand and make use of it; and new information coming from research institutions must reach farmers before it can have any effect on pest management practices.

The bridge between farmers and the other levels of decision making is provided in most countries by extension and advisory services. These may be publicly funded extension services run by ministries of agriculture; or private consultancy services which farmers pay for when they seek their advice; or the training and information sections of agricultural chemical and machinery companies. They all perform the vital function of linking those who use the decision tools described in earlier chapters with those who must decide what action to take on their own farms.

This chapter is about those linkages. It is concerned with the techniques that are available for influencing and informing farmers' pest management decisions. It begins by exploring the nature of extension: this is important because the traditional and simplistic view of extension as the taking of new technology from research institutions to farmers has to be challenged. At the core of all extension work are processes of communication: not just the one-way transmission of ideas to farmers but an exchange and dialogue in which farmer and extension worker try to understand one another as a basis for reaching an acceptable decision. Communication often goes wrong in extension: some basic principles of human communication are therefore discussed which can help to understand and improve pest management extension.

There are many techniques that can be used for extension work in pest management. However, they are not all suited to the same situations and purposes. It is therefore important to analyse each situation and then to identify appropriate extension objectives before techniques can be selected and put into practice. After discussing a range of possible objectives, the chapter reviews extension methods under three headings: individual, group and mass methods. The examples referred to throughout the chapter show that different methods can complement one another to reach a particular set of objectives. The concluding section summarizes the chapter in a five-stage procedure for selecting and using extension techniques for pest management.

WHAT IS EXTENSION?

Put a mixed bag of extension officers in a room and ask them to come up with a statement of the role of extension – and they could be there for a long time. Those involved in extension work may see their task in several different ways. If you spread the net and ask research scientists and a cross section of farmers for their views as well, the variety of statements will multiply. They will probably include the following ideas.

Transfer of technology – To quote Benor (in Rivera and Schram, 1987, p. 145): 'The basic function of an extension service is to promote useful and remunerative technological change among farmers, and to keep agricultural research and other rural services well informed of farmers' conditions and needs'.

This transfer is usually seen as a 'top-down' process; a transfer of new technology from researchers, through extension workers[1], to farmers. But it can also be seen as horizontal – the transfer, or sharing, of new ideas and technology between farmers. Many new wheat and rice varieties have been developed by farmers, who have taken varieties from research stations and bred into them characteristics they find desirable (Brammer, 1980). Paul Richards (1985) argues persuasively that extension services should spend more effort on promoting 'horizontal extension' – i.e. finding out what the most successful farmers are doing and helping other farmers to learn from their experience and expertise (cf. Garforth *et al.*, 1988).[2] After all, this is how agricultural technology and practices evolved and spread over the many centuries before formal research and extension services were established.

Provision of information – Information can be seen as a resource, or input, and the role of the extension worker as being to make that resource available to farmers. This view emphasizes that each farm is different and that what might be appropriate for one farmer would perhaps be inappropriate for another. It is up to the farmer to decide what to do on the farm: it is the extension worker's task to make sure the farmer has enough information to make decisions which are appropriate to the particular circumstances of the farm and of the household which depends on it.

---

[1] 'Extension worker' in this chapter refers to anyone whose job description or professional responsibility includes direct contact with farmers to improve pest management in their farms, crops or livestock.
[2] An example of horizontal extension was the distribution, in Thailand in 1988, of information on pesticide made by farmers from local plants. The information came from a survey of traditional plant protection practices in rice in north-east Thailand (Jonjuabson and Hwai-kham, 1991, p. 32).

Giving advice – This could mean giving farmers in an area general advice on a particular enterprise or problem, or advising an individual farmer what should be done on his or her farm.

Problem solving – Here the extension worker starts with a problem that is faced either by the individual farmer, or by many farmers within a certain area, and helps them find a solution to that problem. This view suggests that the 'agenda' for the extension worker is (or should be) set by the farmers in the locality, and not by the decisions of senior management and subject matter specialists in the extension agency.

Education – It may be felt appropriate to help farmers acquire a more 'scientific' understanding of agriculture so that they can make more 'rational' decisions about their use of resources and reach more informed judgements about new technology and practices (e.g. about the importance of good cultural practices, or recommended pruning regimes). In the context of plant protection, we might anticipate the need for farmers to learn about pest life cycles and to handle concepts related to economic thresholds.

Training – Some extension workers will emphasize the importance of teaching farmers specific skills, both cognitive and manual, to enable them to use new technology – and even to use existing technology more efficiently. Spraying equipment, for example, is often poorly maintained and incorrectly callibrated: training in these key areas may do more to increase the cost-effectiveness of pest control than new technology *per se*[3].

Strengthening the organizational base of farmers – Farmers may be prevented from taking full advantage of pest management techniques by social or economic structures rather than by lack of knowledge or information. If so, it may be appropriate to help them overcome those obstacles. This is a view voiced particularly in relation to small-scale farmers. The formation or strengthening of farmers' organizations can help them stand up against the power (political and/or economic) of landlords, traders, moneylenders and other groups. This can be seen as creating **countervailing power**; or as developing an **acquisition system** rather than a **delivery system** (Axinn, in Rivera and Schram, 1987). The relevance of this view of extension was highlighted in the World Conference on Agrarian Reform and Rural Development (WCARRD) in 1979 and the programme initiatives that came from it (especially the Small Farmers Development Programme and, more recently, the People's Participation Programme, sponsored by FAO).

Supplying inputs, credit and technical services – Aerial spraying, loan of knapsack sprayers, subsidized sale of recommended chemicals and distribution of biological control agents may be seen as valid roles of extension within a pest management programme.

In reality, of course, all of these views are valid. Underlying them all is the recognition that extension services seek in various ways to promote change in agriculture, which requires changes in the way farmers use their resources. This may imply the adoption of new methods or technologies in some situations, and if the methods are totally new – as when farmers are being encouraged to change from a calendar spray strategy to a thresh-

---

[3] Members of a farmers' group in Turkey, to which a graduate adviser is attached, explained that they had learned from him how badly their spraying equipment was performing. Testing, repair, adjustment and maintenance had reduced their costs, reduced the amount of chemicals used and increased their cereal and oilseed output through more effective pest control (discussions with the author, April 1992).

old-based decision strategy or to an integrated pest management approach – this is likely to require education and training. The above views are therefore not mutually exclusive: all are relevant to extension for pest management, as we shall see later.

We can sum up the discussion so far by pointing to three broad interpretations of extension, each of which gives important insights into the activities an extension worker might be engaged in and the objectives he or she is working towards. As a process of **technology transfer**, extension's immediate concern is to bring new (tested) technologies to the attention of those for whom they are considered suitable, to provide enough information and guidance to try them out if they want to, and to encourage them to do so. As a process of **institution building**, extension is concerned with the successful formation of groups of rural people which can then act as a mechanism for self-sustaining development as well as for the diffusion of information and technology. As a process of **dialogue with farmers**, extension's role is to help identify and articulate farmers' problems, potential and technology requirements and then to help them realize their potential for change. In all of this work, an extension agency is helping farmers, both individually and collectively, to reach decisions, find solutions to problems and take action.

A common denominator in these interpretations is communication. All extension work involves communication, whether this is seen as a one-way conveying of information or as an interactive process of sharing ideas in order to reach some common understanding or an agreed decision. Extension workers are, above all else, communicators. To function effectively in that capacity, they need to be aware of some basic facts of life about human communication. These are illustrated most clearly when things go wrong. Here are two examples.

Case 1 – The Ministry of Agriculture in a country in southern Africa regularly mounts educational campaigns on livestock parasites. A slide–tape presentation was commissioned on the subject, from the Ministry's media unit. The media unit staff, all former field extension workers, set to work and duly produced the set of slides, synchronized with a commentary recorded on audiocassette. The slide–tape began by briefly listing the harmful effects of parasites on livestock health and on family, rural and national economies. It then divided the topic into two: external parasites and internal parasites. Under the 'external' heading, various categories of organism were listed, life cycles described and control measures suggested. A similar presentation was given for internal parasites. When the slide–tape was tested for comprehension by livestock owners, the result was – confusion! What is a parasite? Why was the same word being used to talk about both ticks and worms? What's wrong with the farmers' present control measures? Why talk about so many different kinds of worms, each with different treatments? How are we supposed to remember all the information?

Case 2 – A widely distributed poster on safety in handling pesticides includes a picture showing someone filling a drum, labelled 'pesticide', with water from a tap; and another person putting maize cobs into a sack labelled 'pesticide'. Over both these pictures, a red diagonal cross had been drawn. This was intended to mean 'never use empty pesticide containers for storing water or food'. Some farmers, however, thought it meant that empty containers should be washed out before using them again; and that it is a good idea to add a bit of pesticide to maize that is going to be stored, to prevent damage by insects. They saw the red cross as indicating approval — just like it does in election posters when we are being urged to vote for a particular candidate.

So what had gone wrong? In the case of the slide–tape presentation, the media unit had not taken into account the farmers' existing knowledge about parasites, nor the fact that they already take some action against them. More serious was the assumption that farmers would understand the word used for 'parasite', and that in the farmers' view of the world, it is reasonable to link ticks and worms under the same general heading. But in the conceptual framework that the livestock owners used to make sense of the world around them, there was no link between these two. The whole presentation had been structured around a textbook treatment of the topic, beginning with the taxonomy of parasites, rather than being structured around the farmers' perceptions, understanding and priorities.

With the poster, a sign was misinterpreted. This brings us to three basic facts of life about human communication, which apply equally to verbal communication between two people and to communication through print or audio-visual media. Extension workers and those who produce information and training materials on pest management need to take account of these in all their communication activities.

1. We communicate by exchanging 'signs'. We cannot transfer an idea we want to share with someone directly into their minds: we speak to them, we write words on paper, we smile, we show them a picture. All of these are 'signs' through which we expect the other person to understand what we are intending to say. But the trouble with most signs is that they are ambiguous. This is most obvious, perhaps, with symbols such as the red cross in the pesticide poster: a cross can mean many different things, depending on the context in which it is used and the intentions of the person using it. There is therefore every chance that someone will fail to give the sign the intended meaning. The sign itself has no inherent meaning: it is given meaning by the person who uses it to try to convey an idea, and also by the person to whom the idea is being conveyed. There is no guarantee that the two people will attach the same meaning to the one sign.

The same is true of all signs. The word 'pest' can mean several different things; and even in the context of plant protection, can be interpreted in different ways. To some, it really means 'harmful insect', while to others it can cover any organism which is seen as a threat to a plant or animal of economic value. Two people are more likely to attach the same meaning to a sign if they come from the same social, economic, cultural and educational background. In the communication between extension workers and clients, there is often a wide gap of experience and background. Misunderstanding is therefore common and to be expected.

2. Dialogue is a more effective way of achieving most communication objectives than a simple one-way transmission of information. This is true whether our main objective is to give information, or to acquire information. When we communicate with someone else, we often want to share an idea or some information with the other person. We want them to learn something, to acquire new knowledge, to be better informed, to develop a new skill. We often forget, however, that the other person already has a lot of knowledge, skill and information. Only by listening to the other person will we begin to find out what they already know. As the dialogue unfolds, each person will see how the other has reacted to what has been said: have they understood? have they believed? is their reaction favourable or not? what attitudes do they seem to have towards what each other is saying? This feedback is useful: it allows us to adjust the next stages in the dialogue. We may decide to repeat something, to express it in different words, to try to convince the other person with another argument.

Extension workers sometimes have difficulty with the notion of dialogue. They have a thorough training in agriculture and the science that lies behind it. They have a lot of useful information to impart. However, it is vital that they also recognize the usefulness and validity of farmers' own knowledge; that they are prepared to learn as well as to teach; to receive as well as to give information. Extension workers must begin their communication activity from where their clients are: this means engaging in dialogue to find out what they know, and then to build their new information onto what is already there. Listening to farmers is just as vital for the extension worker as talking to them.

The importance of understanding farmers' existing knowledge applies both to the **content** of their knowledge (what they know, think they know and don't know) and to the **conceptual framework** within which that knowledge is available to them. A farmer may recognize a spider or a ladybird, but may not have a concept of predator, or beneficial insect, to enable him or her to make sense of advice on how not to kill them, or even to make sense of the information that these species are useful. Similarly, a farmer may recognize and name a particular plant species on his or her farm, and know a lot about it (including things which the extension worker does not know); but while the extension worker may classify it as a 'weed', the farmer may use quite different words and categories to describe it, indicating that the significance of the plant is quite different to him or her. So when the extension worker speaks of 'weed control', the farmer may not recognize the activity or recommendation as applying to this particular species. Extension workers must be particularly careful here: research suggests (e.g. Richards, 1985) that extension workers – even those brought up in rural areas – have acquired scientists' conceptual frameworks through their education and training which hinder them in their communication with farmers.

There are two points here: first, extension workers should as far as possible avoid using words and phrases which are not familiar or may be misunderstood by farmers; and second, they can use farmers' own concepts to put across their information. There will be times, however, when it is entirely appropriate to introduce new concepts: even here, the principle of being sensitive to what farmers know and understand is important.

Case 3 – A crop protection calendar in an Asian country recommended the timing of preventive and control measures in terms of the number of 'days after emergence'. This was widely misunderstood by farmers as 'days after planting'. This latter term made more sense to farmers: planting is an activity they control, and they know precisely when they did it. They can mark it in the calendar and count the number of days from that date. 'Date of emergence' is, from the scientist's point of view, more accurate as a basis for plant protection recommendations because germination will vary with the climatic conditions. However, it is far less accurate if farmers misunderstand it! In subsequent extension materials, the timing of recommendations was expressed in terms of 'days after planting'.

3. Human memory is fallible. People forget – even more, they mis-remember. We need to be reminded of relevant information, even if we already 'know' it. Indeed, posters and leaflets with technical information, or with advice on the safe use of agrochemicals, may be more important as **reminders** of what farmers have already learned than as initial conveyers of information.

EXTENSION OBJECTIVES FOR PEST MANAGEMENT

The strategic aim of extension is to secure beneficial changes in agriculture. This can be achieved only indirectly, by influencing the decisions and behaviour of farmers. For it is farmers who are the key decision makers in agriculture, and therefore in pest management. They decide what crops to grow and what action to take against pests. The aim of extension is therefore more appropriately seen as to help farmers make sound, well-informed decisions on their holdings. As Röling puts it, the role of extension is to 'induce change in voluntary behaviours' (1988, p. 49).

Fulfilling this aim in relation to pest management presents three particular difficulties to extension workers. First, decision tools which will improve farm level pest management are complex, often requiring new cognitive skills as well as new factual and procedural knowledge. Threshold-based spraying, for example, requires a basic ability to work out projected costs and returns, which in turn demands an understanding of the economics of crop production beyond that which a farmer has needed previously. Without this, no amount of teaching the procedures of sampling and pest recognition will help the farmer make appropriate decisions. In the case of still more complex practices, the difficulty is multiplied. Goodell (1984) showed that in order to apply IPM practices in rice recommended by IRRI, a farmer needed to work through nine or ten steps, each with between two and twelve possible choices, to reach a decision on what action to take.

Second, extension workers are often faced with having to contradict earlier advice, as with a move from calendar- to threshold-based spraying, or with the removal of a particular product from the list of recommended or permitted pest control chemicals. This can have a damaging effect on the credibility of extension workers.

Third, effective pest management, especially in smallholder agriculture, may require collective activity. Most agricultural extension focuses on individual holdings, with extension workers encouraging farmers to adopt changes on their own farms which will benefit them whatever their neighbours choose to do. But many pests do not conveniently confine their attention to a single holding. They can fly, hop, walk, swim or be carried by a host from one to another. This problem is at its most extreme in communal grazing systems, where parasite control can only be effective where all those with cattle grazing a particular area treat them against parasites. Rat control among arable crops is another case where collective action is the only way of achieving a significant degree of control. Extension workers therefore find themselves trying to promote collective decision making and action.

These three difficulties highlight the fact that extension for pest management cannot be reduced to the provision of information and the encouragement of individual farmers to adopt new practices.

Defining the role of extension as helping farmers make (and carry out) sound and well-informed decisions brings into clearer focus the different views of extension with which this chapter began. These views can be seen as a range of possible intermediate objectives which may have to be achieved before a farmer can make a decision and carry it out on the farm. In two different contexts, the same aim may require different sets of objectives. An extension worker may feel, for example, that the most appropriate solution to a particular soil-borne pest may be to fumigate the soil. He or she wants to help farmers decide whether or not to select that course of action. In one area, where farmers are already familiar with the life cycle of similar organisms and have successfully used fumigation before, the extension worker's objective may simply be to inform the farmer about the most suitable chemicals from which to choose, the timing and rate of application, and any special safety precautions. In another area, where farmers' knowledge of soil-borne pests and of fumigation is limited, more relevant objectives might be for farmers to learn how the pest survives in the soil, to be convinced that fumigation is a feasible, effective and economic treatment, and to receive training in how to carry it out safely.

Based on their analysis of the situation, including farmers' circumstances and knowledge, extension workers need to decide on an appropriate objective, and then to select an extension method to achieve it. This does not mean that the decision is predetermined. The extension worker may have views on what decision would be in the best interests of the farmer; in other cases, the extension worker may feel neutral among a range of possible decisions. But in the end, it is the farmer who will decide. The extension worker's role is to provide the necessary support for a well-informed decision to be made.

Table 17.1. Examples of extension objectives in pest management

Objective	Examples
Provide education to develop farmers' knowledge and understanding, or teach them new decision rules and criteria	• life cycle of key pests • the concept of 'economic threshold'
Give information to help make immediate decisions, or to store for use in future decision making	• literature on different equipment • lists of recommended chemicals • current pest incidence in the area
Advise what action to take in present circumstances	• check pest and damage level in the crop now
Create awareness of new ideas, products or practices	• concepts of 'predators', 'beneficial insects', 'resistance to disease' • checking pest damage, as a routine activity
Develop cognitive skills needed to handle new kinds of information	• calculation of pest incidence • calculation of chemical application rate
Train farmers in new practical skills, or to carry out existing practices more efficiently	• sampling within a field • calibration of sprayer • preparation and placement of bait
Facilitate formation of groups, for mutual support, economies of scale or more effective action	• 'scouting' groups, to share information on pest levels • rat control groups, for joint action
Encourage farmers' confidence in their ability to adopt new practices	• show that other farmers in similar circumstances now routinely and successfully use threshold-based decision making

If we assume a particular aim — enabling farmers to take economic decisions about plant protection, based on threshold concepts — the range of possible objectives can be illustrated as shown in Table 17.1.

EXTENSION METHODS

Once an extension objective has been decided, a method can be selected to achieve it. Extension methods can conveniently be grouped into three categories: individual, group and mass methods. This section does not cover the full range of extension methods but illustrates how methods within each category are used in pest management extension. Detailed guidelines on the use of the various methods are given in standard texts on agricultural extension (Swanson, 1984; Oakley and Garforth, 1985; van den Ban and Hawkins, 1988). In all methods, extension aids and mass media, ranging from leaflets and flipcharts to computers and video cassettes, can greatly enhance the effectiveness of the extension worker or adviser's work.

Individual methods

These individual methods bring the extension worker into contact with a single farmer, in the context of a specific problem faced by the farmer at that time, or of a more general discussion of the options available to the farmer for pest management. The contact may take place on the farm or in the office or elsewhere. These methods are relatively expensive in their use of the most costly extension resource — extension workers themselves; but they do allow them to focus their attention and expertise on the specific problems and opportunities facing the individual farmer and can therefore be particularly effective in helping farmers find an appropriate solution to the problem or question they face.

Visits by extension workers to the farm, or by farmers to the office, are excellent opportunities for:

- Information exchange between extension worker and farmer: a dialogue, in which each learns from the other.
- A whole-farm analysis of opportunities and problems.
- Identifying the precise nature of a pest management problem.
- One-to-one training in cognitive or practical skills.

On the farm, the extension worker can quickly diagnose a pest or disease problem, or collect samples to take to a research institution for analysis and identification. In the office, the extension worker may have access to resources such as reference books, equipment manufacturers' literature, posters, preserved specimens of pests and diseased plants, and computer programs for estimating economic returns to expenditure on pest management.

To take full advantage of the benefits of individual extension methods, extension workers must be supported by:

- Transport, both for scheduled farm visits (as in the Training and Visit system), and to respond to an urgent request to go to a farm.

- Up-to-date literature.
- Specialist back-up services including laboratory analysis and expert advice.
- Appropriate extension aids. Apart from their functional use as sources of information, or as teaching aids, these add to the appearance of the office, and show that extension workers take care to keep themselves well informed.

Extension workers cannot be in two places at once. They can, however, increase their accessibility to farmers by informing them when and where they can be contacted. Some keep regular days for office work, so that farmers know they will find them there on those days. If these days coincide with market days, when farmers are likely to come to the village or town where the extension worker is based, so much the better. Planning and publicizing a programme of farm visits and other activities, so that farmers know in advance when the extension worker will be in their locality, also helps. In the Training and Visit system, this is formalized in a regular fortnightly or monthly schedule.

Group methods

These methods are those in which an extension worker is in contact with several farmers at the same time. Contact is usually face-to-face, but may also be through a group newsletter or notice-board. The use of a group method does not imply that there is an established group of farmers, which has an identity as a group and which meets together regularly. Extension can certainly be particularly effective where such groups exist, but most of the methods included in this category do not rely on them. The 'group' element simply refers to the fact that several individuals are involved in the same extension activity.

These methods range from formal, carefully planned events such as field days, study tours and method demonstrations, to informal discussions with no pre-planned agenda. They are useful where the objective is:

- To reach a collective decision on what action to take in the face of a shared problem or opportunity.
- To train farmers in a cost-effective way (i.e. at lower unit cost than one-to-one training).
- To encourage exchange of views and experience among farmers.
- To listen to feedback from farmers' experience with earlier advice or extension activities.
- To pass on information or advice that is relevant to several farmers more efficiently than through individual contact.

Result and method demonstrations are commonly used in pest management extension, particularly where the subject of the demonstration is relatively new in the area. If farmers can see with their own eyes, and discuss with the extension worker and other farmers what they have seen, they can more easily reach a decision about its effectiveness and feasibility.

A result demonstration is usually prepared on the land of a cooperating farmer, so that the conditions under which other farmers see it is similar to that of their own farms. The result of most pest management practices can only be seen after some time, possibly a whole season. The demonstration must therefore by planned carefully, in good time. This includes: securing the cooperation of the farmer on whose land the demonstration is to be held, making sure all necessary equipment and inputs are in place, checking the farmer knows how and when to carry out the necessary operations, and deciding when other farmers should be invited to the demonstration site.

The demonstration will be more effective if the farmer carries out all operations on the site. He or she is then able to explain to others what was done, which in turn carries greater credibility than if the extension worker has to do the work and give the explanations. The demonstration site is of no use if no one comes to see it. Field days can be held at the site; and sign boards briefly describing what has been done on the plot will help spread awareness of the practice more widely. The topic for demonstration will depend on what is appropriate to the area: examples could include disease-resistant crop varieties, pre-emergence herbicides, and improved accommodation for livestock.

Result demonstrations can also be a way of showing a wider audience what some farmers are already doing. If a farmer, for example, has made some improvements to cattle accommodation, by giving it more ventilation and light and making it easier to keep clean, the extension worker and farmer could agree to use it as a demonstration of the effect of these improvements on the health and productivity of the animals. The demonstration in this case would be facilitating horizontal extension rather than the introduction of a new practice from outside the area.

Some result demonstrations include a control plot or condition, so that the results of the demonstrated practice can be seen next to existing or alternative practices. While this can be visually striking, it is not usually necessary: in most cases farmers are able to evaluate the results of the demonstration, without such a control, against their knowledge of their own plots.

Method demonstrations show farmers how an operation or task is carried out. This is useful when a new practice is being introduced – for example, how to apply herbicide safely and effectively; or when a problem emerges with an existing practice – such as the cleaning and maintenance of sprayers, or handling livestock when dosing them for parasites.

Demonstrating a practice is not easy and is very different from simply carrying out the practice by oneself. While demonstrating, the extension worker must be able to explain clearly what is being done and why; to speak in the direction of the farmers so that they can hear; answer their questions; and make sure they can see what is happening. This is only possible if the extension worker has carefully analysed the practice or task. This involves breaking it down into its component stages, identifying clearly what has to be done in each stage, deciding what points need stressing or repeating, and anticipating what might go wrong when farmers try it on their own farms. It is then necessary to practice the task; and also to practice demonstrating it. Some extension workers find it better to have someone to assist them: a farmer might carry out the task while the extension worker gives the explanation. In this case, preparation and practice with the assistant is vital. During the demonstration, farmers should have an opportunity to try the task for themselves. Encouragement and correction by the extension worker is an important aid to farmers' success in learning the new skill.

A **problem census** has a very different set of objectives from demonstrations. There is no message to give, no skill to teach, no result with which to impress farmers. The census brings together farmers within a village or other unit to analyse pest problems in the area and on their farms. Problems are then prioritized and the group decides what action to take. The role of the extension worker is to encourage the farmers to meet together in the first place and then to facilitate the discussion by offering some structure or agenda and providing technical expertise or information when it is appropriate. The stimulus for undertaking the census may be the extension worker's perception that many farmers face a common problem which cannot be tackled effectively or economically on an individual basis, and that a careful analysis of the problem is needed before appropriate solutions can be found. Or the initiative may come from farmers who see their pest problems increasing and want to develop a coherent strategy for dealing with them.

The basic features of a problem census are straightforward (Hoare, 1987). A group of farmers assembles, representative of the types of farms and enterprises in the village. All those involved in husbandry of growing and stored crops and of animals should be represented so that all points of view on a particular problem can be heard. If women, for example, are generally responsible for weeding or for particular crops, while men are responsible for land preparation or for other crops, it is important that both are well represented. The group then lists the pest problems faced in the village and analyses them: why do they think these problems occur? is it because of intensification of farming? or changes in cropping patterns? or climate? or shortage of labour? The problems are then put in priority order and the group decides how the high priority ones should be tackled.

However, the apparent simplicity is misleading. Whoever is leading the problem census needs skills in handling group discussion. They must make sure that the process is not dominated by one or two influential speakers; help to resolve misunderstandings and arguments between different points of view; summarize the discussion from time to time; keep the discussion from being side-tracked into peripheral issues; and try where possible to achieve consensus. If there are more than 10 to 15 people taking part, some will feel inhibited from contributing. It is helpful then to divide into smaller groups, with each group drawing up its own list and analysis for sharing with the others in a plenary session. Meeting in separate groups is also a good way of ensuring that smaller scale or less influential farmers can express their views without feeling intimidated.

Farmer tours and visits reverse the normal extension procedure of bringing information and advice to the farmer: farmers instead go to where the information and advice are available. Throughout the ages, people travelling from one area to another have brought new ideas about agriculture with them. Extension services in many countries now accelerate this process by organizing visits and tours for farmers. These can be to research institutions, machinery manufacturers, agricultural fairs, university faculties of agriculture, or to farms. The objective is to expose farmers to new ideas which might be applicable to their own area. A group of community leaders might visit another community in the same region which has successfully controlled damage by rats; or a farmers' group contemplating building and equipping a cattle dip might visit a working dip operated by a neighbouring group. Such visits are an opportunity to find out from people who have already implemented a decision how they did it, what problems they faced, how they overcame them and what the benefits have been. The atmosphere of a visit away from their home area can also stimulate discussion among the visitors themselves, helping to mould them as a group that can take a collective decision and remain committed to an idea on their return home.

Visits and tours need careful planning. Once it has been decided that a visit is appropriate, and a location (or a series of locations) has been identified, there are a lot of details to be arranged including travel, accommodation, liaison with host organizations and finance. More important from an extension point of view is the question of who takes part and how the extension worker should follow up the visit once it is over.

A local version of the visit is the **farm walk**, in which an extension worker and a host farmer take a group round the host's farm. This can be useful where the host farmer has made some changes in pest management which might be applicable to others in the area, and where the effects of those changes are clearly visible. A farm walk is therefore a good opportunity:

- To show how the pest situation on a farm has been improved by basic husbandry practices (e.g. soil preparation, seed rate, management of slurry, animal hygiene).
- To discuss why the farmer has made particular pest management decisions in the farm's various enterprises.
- To show the results of earlier pest management decisions, particularly where these are new to the area.

In farm walks, visits and tours, most of the advice and information that farmers obtain is given by the researchers, farmers, company representatives or community leaders whom they meet. The extension worker brings the farmers and the source of information together. This is a useful reminder that extension workers are not the farmers' only source of information. Farmers have many other channels of communication about pest management. **Pesticide retailers**, for example, have much more direct contact with farmers than most extension workers. However, they often do not have sufficient training to be able to give accurate information and advice. The information they do have may be limited to sales and promotional material provided by the companies whose products they sell. Extension agencies can improve the flow of useful, accurate information by:

- Giving posters and leaflets to retailers which will help them and farmers identify pests and the recommended ways of dealing with them.
- Holding training courses and seminars for retailers.
- Giving regular, up to date information to retailers about the current pest situation in the area.

Retailers have a particularly important role to play in spreading safe practice in the storage, handling and application of pest management chemicals. Many farmers have difficulty deciphering the precautions and instructions printed on pesticide containers and accompanying leaflets. Extension workers can encourage retailers in their area to explain these to farmers whenever they buy a product.

More generally, extension workers can make use of existing communication networks by finding out which farmers are regarded by others as sources of information about pest management. Who do farmers go to if they see a problem in their fields or animals? This approach forms part of the Training and Visit system, where extension work is concentrated on selected 'contact farmers' in the expectation that information will pass from them to other farmers in the area. Often, however, contact farmers are selected who are not regarded as useful sources of information by others and so the flow of information is restricted.

In all individual and group extension, the most important tool is the extension worker. Nothing can substitute for extension workers who:

- Keep themselves informed about the latest pest management techniques appropriate to the farming systems in their area.
- Spend more time in the field than the office so that they are continually updating their knowledge of the current pest situation.
- Are prepared to admit (to themselves and to farmers) when their own knowledge is insufficient.
- Develop their own network of expertise – colleagues in their own organization, technical representatives from input supply companies, university scientists – which they can call on for rapid support and advice.

Early identification of a pest problem is a vital first step in preventing it from becoming serious. **Scouting and surveillance systems** can therefore be important tools in the fight against pests. But how can these be used effectively for extension? Where the systems are designed to provide information to senior decision makers within the ministry of agriculture, the extension benefits may be small. A decision may be made to order aerial spraying, or to recommend that farmers take some specified action; but the information is not enhancing farmers' abilities to take appropriate decisions about their own crops. Where fields are sampled as part of a surveillance system, the data should at least be shared with, and explained to, those who farm them. In the long run, a complementary programme to train farmers to survey their own crops and to calculate when to take action may be necessary for the surveillance system to achieve its full potential. This was the case in the Philippines in the early 1980s, where a surveillance system for rice provided regional analyses of pest incidence and dynamics (with a two week time lag) but farm level information only for those farms included in the surveillance sample. Complementary approaches were developed, including training courses at village level, to enable all farmers to make better informed decisions on pest management.

Mass methods

These include all those extension methods which have the ability to reach large numbers of people with the same information in a relatively short period of time. In the case of broadcast mass media (radio and television) and daily newspapers, the information can reach a mass audience almost simultaneously. With other methods (film, video, posters, drama), the period is extended. Unlike group methods, the 'mass' audience is not all in one place: the attraction of mass media is that the production costs are spread among a large, scattered audience. Each member of the audience may see or hear the media separately but the content only has to be prepared once. In the case of broadcast media, distribution is also simple once the transmitters are in place. With print, and video and audio cassettes, distribution is more complex and often leads to delays in getting the media to the audience. In print media, cost reduction can be taken even further by publishing in different languages but using the same illustrations: this has been done by IRRI and CIMMYT, for example, with their pocket guides to pests in rice and in wheat.

Mass media can be used in many ways. They can convey urgent information rapidly, with a minimum of distortion. They can expose farmers and communities to what other farmers and communities are doing to tackle pest problems. They can reinforce and remind farmers of what they have heard from extension workers, retailers and other sources. Visual media, particularly television, video and film, can carry a strong emotional appeal: showing, for example, the devastation that a particular pest can cause if it is not controlled; or, more positively, showing how a farm or an area has been transformed by an effective pest management programme. They can also compress time very dramatically: in five minutes, a television programme can show what happens over a complete season if weed control is not done in time, or if basic hygiene is not followed in livestock husbandry.

Visual media are particularly useful in pest identification. In East Africa, the Desert Locust Control Organization has produced video cassettes to teach extension workers how to recognize armyworm, and to stress the importance of early identification and reporting so that strategic decisions on control measures can be taken. This was done because it became clear that field extension workers were unfamiliar with the pest. However, while video and television can leave a long term general impression, precise information is quickly forgotten. Therefore booklets have also been produced, for extension workers to refer to in the field when trying to identify the pest.

A decision to produce and distribute mass media for pest management extension does not normally rest with the extension worker: it requires strategic planning at a high level. The choice of what media to use will depend on the media facilities already available to the farmers of the area. In an increasing number of countries, television is already widespread in rural areas and is being used by ministries of agriculture for extension and education purposes. In far more countries, radio remains the most accessible mass media channel and is widely used. In some countries, radio or television 'schools' provide systematic education to farmers on improved crop and animal husbandry.

Extension workers can use these channels in two main ways: by providing news and information from the field to producers of mass media, and by using the media that are available in their contacts with farmers. In Thailand and the Philippines, some extension workers with an interest in radio present a regular programme on local radio about the current pest situation. By doing so, they reach far more farmers than they can in a week of individual farm visits. Others write columns for newspapers; or send reports of demonstrations to agricultural broadcasters or newspaper editors.

Print media on pest management are widely available. Unfortunately they are not used in the field as much as they could be. A poster is more effective if it is displayed in a village than on the wall of a government office. It is more effective still if the extension worker uses it in a training session or a discussion. Booklets on pest identification, designed to fit into a pocket, are more use in the hands of the extension worker during a visit to a farmer's field than on the office shelf. Such booklets do not only help the extension worker identify a pest problem: they can be used to teach farmers skills of pest identification.

An advantage of print media is that they can be available whenever they are needed. This is particularly important in pest management, as many decisions have to be taken in the field and precise information is often required which farmers cannot be expected to remember accurately. An extension worker can prepare a handout of the main points covered in a seminar for farmers to refer to later. Stickers have been produced, with reminders on safety precautions or on calibration, for farmers to stick onto their spraying equipment. Pesticide manufacturers and distributors put technical and safety information on the labels of containers, so that it is available to the farmer when the chemical is used. However, such information will only be used if farmers accept its importance, and if they can read and understand it. In the last few years, international conventions have been adopted for the provision of safety information, using a consistent set of symbols to indicate what precautions are necessary for a particular product. Once the set of symbols has been explained to farmers (by a pesticide retailer or extension worker, or in a television programme, for example), they should be able to interpret the particular combination of symbols on a label.

Another approach to designing print material to improve its usefulness to farmers is the algorithm: this begins with a question, to which the farmer gives a 'yes' or 'no' answer, which in turn leads to another question with a 'yes' or 'no' answer, until eventually a conclusion is reached. To be useful, the algorithm should follow the steps of a decision-making procedure that has been shown to give satisfactory and meaningful conclusions (see Chapter 11). An algorithm can be designed, for example, to guide the farmer to a decision on whether to take any pest management action on a particular field; or to an accurate identification of a pest.

Giving farmers maximum exposure to useful information is an important reason for establishing Farmers Centres in rural communities. A range of material can be kept here: posters and leaflets, daily or weekly newspapers, radio and television, and preserved specimens of pests and parts of plants and animals. It is unusual for such centres to be established only for pest management programmes (although it has been done in some Asian countries). Those responsible for pest management extension can, however, make full use of existing centres, and indeed of other local meeting places such as tea houses and rural libraries and information centres. In Thailand, a chemical manufacturer produced video cassettes on pesticide safety — using visual images from a poster that it was also distributing widely — and arranged for these to be played in rural banks so that farmers coming to the bank would be exposed to the information while waiting to carry out their transactions.

It will be clear from this discussion that each medium has its own particular advantages and that media can be more effective if they are used in combination rather than alone. There are usually opportunities for extension workers to use a range of media in their regular work: a farm visit or a demonstration can be a more useful event if supported by carefully chosen visual aids, or handouts for reference. The term 'mass media' should, however, be treated carefully — for two reasons. First, although the production and distribution of the media may be done centrally, the information they contain does not have to come from a central point. Indeed, mass media can be an effective means of sharing the experience of farmers and communities with a wide audience, thus becoming tools of 'horizontal extension'. Second, the audience may be large but it will not be homogeneous. The larger the audience, the more difficult it is to design content that will be relevant and comprehensible to all or most of its members. Mass media producers need to be just as aware of the principles of human communication as field level extension workers. They need to find out what their audience knows and wants to know; and to check whether their media content is understood. This requires a commitment to audience research and feedback. For media which will have a long life (such as most print media and video cassettes), pre-testing will identify weaknesses which can be put right before final production and distribution. It is better to spend a bit of time and money on pre-testing and revision, than distribute large quantities of media which are misinterpreted or misleading (see Case 4). With more ephemeral media, such as daily farm broadcasts, audience feedback and evaluation will guide future decisions on content and format.

Case 4 – A poster describing the life cycle of the tape worm, and its transmission between cattle and humans, was tested in a country in southern Africa. Four illustrations depicting different stages (preparing and eating meat, defecating, cattle grazing, meat on the butchers' stall) were connected by arrows. Most of those who were shown the poster correctly identified what was happening in the four illustrations. The meaning of the arrows, however, was not clear. Some said they were footpaths through the bush; others saw them as snakes; others as signposts. Nobody interpreted them as the designer had intended.

In a strategic campaign, or a long-term pest management programme, several different extension methods will be used, each selected to meet specific objectives within the overall programme. A well-documented example of this is a series of campaigns on rodent control in Bangladesh, undertaken by the Bangladesh government with technical support from FAO and others (Adhikarya and Posamentier, 1987).[4] Broadcast mass media were used to create awareness of the scale of the problem and of successful efforts by communities to tackle it; print media reminded people of baiting techniques and precautions in using rodenticides; method demonstrations were held by extension workers; training programmes were arranged at village level; popular support for the campaign was encouraged through community leaders; each year, the campaign organizers reviewed experience and adjusted the programme for the next year.

CONCLUSION

Decisions on extension methods are made at different levels – from the daily decisions of the field level worker to strategic decisions of a Ministry of Agriculture. At all levels, there is a wide range of options when it comes to choosing what methods to use. The choice can only be made when the extension objectives are clear. These in turn should be based on a clear understanding of the situation in which pest management extension is going to be carried out. We can summarize the discussion of this chapter by suggesting a five-stage procedure.

Stage 1: Research phase
  - find out what farmers know about the situation;
  - collate available technical and scientific information;
  - clarify the situation, problems and solutions.

Stage 2: Decide extension objectives.

Stage 3: Select extension methods that are
  - appropriate to the farmers and to the extension organization;

---

[4] A useful summary of the campaigns is given in Adhikarya, 1989.

- best suited to meeting the objectives;
- complementary and reinforcing.

Stage 4: Plan and implement extension activities

Stage 5: Monitor and evaluate
- the implementation (are the activites taking place as planned?);
- the results (are the planned activities having the results that were anticipated?).

In the case of a single extension activity planned and carried out by a field level worker, these stages may follow the strict sequence, with each stage completed before the next one begins. With a longer-term or more complex programme, however, the stages will overlap: preliminary decisions on extension methods may reveal the need for further research; while monitoring should be a continuous process so that the programme can be adjusted in the light of progress in the field. In a long-term programme, the five stages become a regular cycle, with evaluation forming part of the research stage of the next phase of activity.

REFERENCES

Adhikarya, R. (1989) The strategic extension campaigns on rat control in Bangladesh. In: Rice and Atkin, pp. 230–232.

Adhikarya, R. and Posamentier, H. (1987) *Motivating Farmers for Action: How Strategic Multi-Media Campaigns can Help.* GTZ, Eschborn, Federal Republic of Germany.

Axinn, G.H. (1987) The different systems of agricultural extension education with special attention to Asia and Africa. In: Rivera and Schram, pp. 103–114.

Benor, D. (1987) Training and visit extension: Back to basics. In: Rivera and Schram, pp. 137–148.

Brammer, H. (1980) Some innovations don't wait for experts: a report on applied research by Bangladesh peasants. *Ceres* 13, 24–28.

Garforth, C., van Schoot, C. and Maarse, L. (1988) The role of extension in developing the use of rangelands. *Agricultural Administration and Extension* 30, 325–334.

Goodell, G. (1984) Challenges to integrated pest management research and extension in the Third World: Do we really want IPM to work? *Bulletin Entomological Society of America* 30, 18–26.

Hoare, P. (1987) Thailand: extension and equity in highland rainfed agriculture. *Reading Rural Development Communications Bulletin* 22, 33–39.

Jonjuabson, L. and Hwai-kham, A. (1991) A summary of some Thai NGO experiences in sustainable agriculture in collaboration with government agencies. In: *Agricultural Administration (Research and Extension) Network Paper No 28.* Overseas Development Institute, London, pp. 24–35.

Oakley, P. and Garforth, C. (1985) *A Guide to Extension Training.* FAO, Rome.

Rice, R. E. and Atkin, C.K. (eds) (1989) *Public Communication Campaigns.* Second edition. Sage, Newbury Park and London.

Richards, P. (1985) *Indigenous Agricultural Revolution.* Hutchinson, London.

Rivera, W.M. and Schram, S. G. (eds) (1987) *Agricultural Extension Worldwide.* Croom Helm, London.

Röling, N. (1988) *Extension Science: Information Systems in Agricultural Development.* Cambridge University Press, Cambridge.

Swanson, B. (ed.) (1984) *Agricultural Extension: A Manual.* FAO, Rome.

van den Ban, A. and Hawkins, S. (1988) *Agricultural Extension.* Longman Scientific and Technical, Harlow, UK.

Index

Page numbers in *italics* refer to figures and tables.